Chemistry and the Living Organism

Sixth Edition

Molly M. Bloomfield   *Oregon State University*

Lawrence J. Stephens   *Elmira College*

John Wiley & Sons, Inc.

New York • Chichester • Brisbane • Toronto • Singapore

ISBN 0-471-12078-2

Printed in the United States of America

10 9 8 7 6 5 4

Preface

The organization of this study guide to the sixth edition of CHEMISTRY AND THE LIVING ORGANISM is designed to complement the textbook. The following is a list of important features of the study guide:

Each chapter begins with a summary of the important concepts covered, and then has a section-by-section discussion of these topics. Emphasis is placed on those topics that students often find most difficult. A list of important terms is included for each section.

Numerous example problems are included. These problems focus on areas in which students often need additional help. The *STEP* problem solving strategy is used to solve these examples. This method is called *STEP* because you are asked to:

*S*ee the question:	Analyze the problem to determine what is being asked.
*T*hink it through:	Based on the information in the problem and other facts you know, choose a method for solving the problem.
*E*xecute the math:	Do the mathematical calculations necessary to arrive at the answer.
*P*repare the answer	Check the answer to be certain that it has the required units of measure and the correct number of significant digits.

Each chapter contains many self-test questions, called "Check Your Understanding," to help the student measure her or his comprehension of the topics covered. The answers to these problems are found at the end of each chapter.

"To The Student" provides helpful tips on how to get the most benefit from the study guide. The emphasis is on a systematic approach to the study of chemistry using the text, this study guide and class notes.

Appendix I contains the answers to the even-numbered end-of-chapter problems in the text. Answers to the odd-numbered end-of-chapter problems and the Check Your Understanding exercises in the text are found in Appendix 3 in the text.

To The Student

This study guide accompanies CHEMISTRY AND THE LIVING ORGANISM, Sixth Edition. We wrote it to help you better understand the principles presented in the text. The study guide will help you in picking out the important concepts in each chapter, and will aid your review when studying for exams.

The study guide begins with a mathematics review that will assist those of you whose math might be a bit rusty. This review reinforces the material found in Appendices 1 and 2 and in Sections 1.7 through 1.10 in your text.

The study guide closely follows the topics in the text. You can use both books together to increase the effectiveness of your study time. Each chapter in the study guide begins with a paragraph summarizing the topics covered in that chapter. If you read this summary before reading the chapter in the text, it will help you to identify the important material to be covered. Chapters are divided into sections matching the sections in the text.

Quickly read through a complete chapter in the text for the first time, then begin to read it again section-by-section. As you do, also read the section summaries in the study guide. Make your reading of the material an active process. Take notes, outline or underline as you read. Write out a definition in your own words for each of the important terms listed in the guide. To help you with these definitions, there is a Glossary of all important terms at the end of your text.

Be systematic in your approach to problem solving. The *STEP* strategy provides just such a system. This method is used in the example problems in this study guide and throughout the text. If you use the *STEP* strategy, you will find that many problems are not as difficult as they look. Check your understanding of the material by answering the *Check Your Understanding* questions in the study guide. The answers to all these questions are found at the end of each chapter in the guide. In addition, the answers to the even-numbered problems in the text have been included at the end of the study guide. Your class notes and this study guide should provide excellent review materials for quizzes and exams.

We hope this study guide will make your study of chemistry more understandable and enjoyable.

Molly M. Bloomfield

Lawrence J. Stephens

Contents

Introduction   MATHEMATICS REVIEW

In this chemistry course, you will often be asked to solve numerical problems. This math review describes the basic mathematical principles you must know to solve such problems. You will find it helpful to study this section carefully before beginning the exercises in the Study Guide. You might want to review it again before beginning Chapters 3, 6, 10, and 11 in the text. This math review is meant to go along with the material covered in Sections 1.7-1.10 and Appendices 1 and 2 in the text.

Numbers in Chemistry
You will find that the numbers used in chemical calculations are expressed not only as whole numbers (1, 8, 295), but also as fractions (8/11, 194/278), decimals (8.45, 413.8), or numbers in exponential form (8×10^{-2}, 2.37×10^{12}). Most numbers will be positive, but a few will be negative numbers (-8, -23.67) with values less than zero. The numbers used in chemical calculations are usually measurements. Therefore, they should have some unit of measurement attached to them so that we know what is being measured (8.45 centimeters, 478 grams). Pay special attention to the use of the correct number of significant figures. Significant figures indicate the degree of precision of a measurement. We should never indicate that a measurement is more precise than it actually is.

Fractions and Decimals
Fractions can be converted into decimal numbers by performing the division indicated by the fraction. That is, 3/5 actually means "3 divided by 5", and if we divide 5 into 3 we get the answer 0.6. Thus, the fraction 3/5 is the same as the decimal number 0.6.

Numbers Raised to Powers
When a number is multiplied by itself, that number is said to be squared. This is represented by writing a superscript of 2 above and to the right of the number:

$$6 \times 6 = 6^2 = 36$$

To cube a number means to multiply that number by itself 3 times:

$$(1.3)^3 = (1.3) \times (1.3) \times (1.3) = 2.2 \ (2.197 \text{ calculator answer})$$

The power of a number tells us how many times the number is to be multiplied times itself: 1.3 to the power 3 (1.3 to the third) equals 2.2.

Numbers in Exponential Form

Appendix 1 in the text explains how to write numbers in exponential form. Read this Appendix carefully, and then try the following problems.

Check Your Understanding _____

For questions 1 to 10, write the number in exponential form.
1. 1560 1.56×10^3
2. 7,432,000 7.432×10^6
3. 0.0046 4.6×10^{-3}
4. 145 1.45×10^2
5. 0.025 2.5×10^{-2}
6. 0.0000007 7×10^{-7}
7. 456,000 4.56×10^5
8. 0.000041 4.1×10^{-5}
9. 25,000 2.5×10^4
10. 0.15 1.5×10^{-1}

If you had difficulty with these problems, study the following discussion and then try them again.

A number expressed in exponential form is written as a number between 1 and 10 (called the coefficient), multiplied by the number 10 raised to some power. Consider, for example, the number 2165. You know that this number has a value that is just a little greater than 2 times the number 1000. But the number 1000 is the number 10 to the third power:

$$1000 = 10 \times 10 \times 10 = 10^3$$

Thus, 2165 is just a little more than 2 times 10 to the third; specifically, $2165 = 2.165 \times 10^3$. Notice that to change 2165 to 2.165 we had to move the decimal point 3 positions to the left; however, each of these decimal place moves corresponds to changing the number by a power of ten. We must multiply by ten for each place that we move the decimal point to the left, so that our final answer will have the same numerical value as 2165. Thus, "three moves to the left" requires that we multiply our final number by $10 \times 10 \times 10$ or 10^3.

Now consider the number 0.000158. When this number is written in exponential form, the coefficient is 1.58. To change 0.000158 to 1.58, the decimal point is moved 4 places to the right. Each of these decimal moves corresponds to increasing the number by a power of ten. Thus, we must divide the number by $10 \times 10 \times 10 \times 10$ or 10^4. Dividing by 10^4 is the same as multiplying by 10^{-4}, so the number is written, 1.58×10^{-4}.

From these examples we can state simple rules for changing numbers into their exponential form:

1.   First, count the number of places that the decimal point must be moved to create a coefficient between 1 and 10.

246,000	2.46000	2.46	5 places to the left
0.00485	4.85	4.85	3 places to the right

2. The power of 10 to use is equal to the number of places that the decimal point was moved. However, if the original number was less than 1.0 (and it was necessary to move the decimal point to the right), a minus sign is written in front of the power of 10.

$$246,000 = 2.46 \times 10^5$$
$$0.00485 = 4.85 \times 10^{-3}$$

Now try the *Check Your Understanding* questions 1 to 10 again, following these rules.

Mathematical Operations

Addition and Subtraction

You will rarely need to add or subtract fractions in chemical calculations, but you will be required to add or subtract numbers expressed as decimals or in exponential form. No matter in what form the numbers appear, however, the most important thing to check is that you are adding or subtracting numbers that represent exactly the same type of units. For example, it makes no sense to add 6 inches to 3 feet and to claim that you end up with $6 + 3 = 9$ of something. Either you must express both measurements in inches (6 inches + 36 inches = 42 inches) or in feet (0.5 feet + 3 feet = 3.5 feet). Section 1.9 describes how to use conversion factors to convert numbers so that they all represent the same kind of units.

Decimal Numbers. To add or subtract numbers expressed as decimals, the only rule is to line up all the decimal points one under the other before you start adding or subtracting. The decimal point in the answer will appear under all the other decimal points.

Example ───

1. Add the following numbers: 3.4, 0.026, 274.35, 10.458, and 0.14.

    Lining up all the decimal points one under the other and then add the numbers.

$$
\begin{array}{r}
3.4 \\
0.026 \\
274.35 \\
10.458 \\
0.14 \\
\hline
288.374
\end{array}
$$

The least precise number has one digit after the decimal point, the answer is 288.4

2.   Subtract 10.476 from 283.9 .

Lining up the decimal points, we have

$$283.9$$
$$-10.476$$

Because adding zeros to the numbers on the right of the decimal point doesn't change the number, we can rewrite the problem in a form that is easier to work with. (Remember that these added zeros are not, in this case, significant figures)

$$283.900$$
$$-10.476$$
$$273.424$$

The answer to the correct number of significant figures is 273.4

Numbers in Exponential Form. To add or subtract numbers in exponential form, all numbers must first be written to the same power of 10 (that is, all the exponents of the number 10 must be the same). If you are unsure of how to perform this operation, see the section in this math review that discusses multiplication of numbers in exponential form. Once all the numbers are written to the same power of 10, the coefficients of the numbers are added or subtracted according to the procedure we just discussed for decimal numbers.

Example _____

1.   Add 6.29×10^4, 1.79×10^6, and 25.77×10^4.

First rewrite the numbers so that they all have the same power of ten. Next, line up the decimal points and add zeros to the right of the decimal point. Then add the coefficients and, finally, round the answer to the correct number of significant figures.

One way to do this is to rewrite

$$1.79 \times 10^6 = 1.79 \times 10^2 \times 10^4 = 179 \times 10^4$$

Then add the coefficients
$$6.29 \times 10^4$$
$$179.00 \times 10^4$$
$$25.77 \times 10^4$$
$$211.06 \times 10^4$$

The answer, to the correct number of significant figures, is 211 X 10^4. In correct exponential form it is 2.11 X 10^6.

2.   Add 1.043×10^3 to 5.62×10^3.

Because both numbers have the same power of 10, we can just add the two numbers together (being sure to line up the decimal points).

$$\begin{array}{r} 1.043 \times\ 10^3 \\ 5.62\ \ \times\ 10^3 \\ \hline 6.663 \times\ 10^3 \ (6.66\ X\ 10^3) \end{array}$$

3.   Subtract 7.47×10^{-5} from 2.16×10^{-4}.

First, rewrite $2.16 \times 10^{-4} = 2.16 \times 10 \times 10^{-5} = 21.6 \times 10^{-5}$.
Then,

$$\begin{array}{r} 21.60 \times 10^{-5} \\ -\ 7.47 \times 10^{-5} \\ \hline 14.13 \times 10^{-5}\ (14.1\ X\ 10^{-5}) \end{array}$$

Check Your Understanding ─────────────────────────────

For questions 11 to 20, perform the indicated calculations.

11.	$25.4 + 0.35 + 5.3$	16.	$145 - 0.136$
12.	$0.1 + 0.004 + 0.0005$	17.	$6.485 - 0.01$
13.	$320.1 \times 10^6 + 0.02 \times 10^9$	18.	$48.0 \times 10^6 - 360 \times 10^5$
14.	$46.03 + 140 + 0.2 + 0.0015$	19.	$1.2 \times 10^{-8} - 60 \times 10^{-6}$
15.	$45 \times 10^{-5} + 3.6 \times 10^{-3}$	20.	$6 \times 10^{-8} - 0.45 \times 10^{-7}$

Multiplication
The multiplication of a number "a" times another number "b" can be expressed in several ways:

$$a \times b,\ \ a(b),\ \ (a)(b),\ \ \text{or} \quad \begin{array}{r} a \\ x\ b \end{array}$$

Fractions. When several fractions are multiplied together, the result will be another fraction whose top part (or numerator) is the product of all the other numerators multiplied together, and whose bottom part (or denominator) is the product of all the other denominators multiplied together. If the various numbers that appear in the fractions also have units of measure attached to them, the units of measure are also multiplied together to form a "fraction" made up of various units of measure. These units will cancel each other out if they appear in both the numerator and the denominator.

Example

1. Multiply $\frac{4}{5}$ times $\frac{3}{7}$.

$$\frac{4}{5} \times \frac{3}{7} = \frac{4 \times 3}{5 \times 7} = \frac{12}{35}$$

2. Multiply together $\frac{1}{5}$, $\frac{2}{3}$, $\frac{7}{12}$ and $\frac{5}{8}$

$$\frac{1}{5} \times \frac{2}{3} \times \frac{7}{12} \times \frac{5}{8} = \frac{1 \times 2 \times 7 \times 5}{5 \times 3 \times 12 \times 8} = \frac{70}{1440} = \frac{7}{144}$$

3. $15.0 \text{ yd} \times \frac{3 \text{ ft}}{1 \text{ yd}} \times \frac{12 \text{ in}}{1 \text{ ft}} = \frac{15 \times 3 \times 12}{1 \times 1} = \frac{\text{yd} \times \text{ft} \times \text{in}}{\text{yd} \times \text{ft}} = 540 \text{ in}$

4. $\frac{6.3 \text{ grams}}{3.0 \text{ moles}} \times \frac{2.0 \text{ moles}}{1.0 \text{ liter}} \times \frac{1 \text{ liter}}{1000 \text{ milliliters}}$

$$= \left(\frac{6.3 \times 2.0 \times 1}{3 \times 1 \quad 1000} \right) \left(\frac{\text{grams} \times \text{moles} \times \text{liter}}{\text{moles} \times \text{liter} \times \text{milliliter}} \right)$$

$$= \frac{12.6 \text{ grams}}{3000 \text{ milliliters}} = 0.0042 \frac{\text{grams}}{\text{milliliters}}$$

Decimal Numbers. When several decimal numbers are multiplied together, the result is another decimal number. If the decimal numbers also have units of measure attached to them, the units of measure are multiplied together just as was the case for fractions.

Example

1. Multiply 2.12 times 4.731.

$$\begin{array}{r} 2.12 \\ \times\ 4.731 \\ \hline 10.02972 \ (10.0) \end{array}$$

2. What is the area of a carpet that measures 4.2 ft by 3.5 ft?

$$4.2 \text{ ft} \times 3.5 \text{ ft} = 4.2 \times 3.5 \times \text{ft} \times \text{ft}$$

$$= 14.70 \text{ ft}^2 \text{ (square feet)} (15 \text{ ft}^2)$$

3. What is the volume of a box that measures 2.4 cm by 1.58 cm by 10.2 cm?

$$2.4 \text{ cm} \times 1.58 \text{ cm} \times 10.2 \text{ cm} = 2.4 \times 1.58 \times 10.2 \times \text{cm} \times \text{cm} \times \text{cm}$$

$$= 38.6784 \text{ cm}^3 \text{ (cubic centimeters)} (39 \text{ cm}^3)$$

Numbers in Exponential Form. For a review of the multiplication of numbers in exponential form, reread Appendix 1 in the textbook. Rewriting a number in exponential form so that it appears with a different power of 10 is just an application of the rules for multiplication. For example, suppose we wanted to rewrite the number 4.86×10^5 as a number times 10^2.
To do this, we would follow these steps:

1. Rewrite the original power of 10 as two powers of 10 multiplied times each other, with one of the two powers of 10 being the power that we want to end up with. In this example, we would write

$$10^5 = 10^{(3 + 2)} = 10^3 \times 10^2$$

2. Then multiply the coefficient by the power of 10 that we are not interested in, leaving behind the power of 10 that we wanted. In our example,

$$4.86 \times 10^5 = 4.86 \times 10^3 \times 10^2 = 4860 \times 10^2$$

Example _____

1. Rewrite the number 8.2×10^{-6} as a number times 10^{-4}.

$$8.2 \times 10^{-6} = 8.2 \times 10^{[-2 + (-4)]} = 8.2 \times 10^{-2} \times 10^{-4}$$
$$= 0.082 \times 10^{-4}$$

2. Rewrite the number 537.9×10^3 as a number whose coefficient is between 1 and 10.

Using the fact that $537.9 = 5.379 \times 10^2$, we have

$$537.9 \times 10^3 = 5.379 \times 10^2 \times 10^3 = 5.379 \times 10^{(2+3)}$$
$$= 5.379 \times 10^5$$

Check Your Understanding

For questions 21 to 29, perform the indicated multiplication:

21.

$$\frac{3}{7} \; X \; \frac{22}{12} \; X \; \frac{4}{5}$$

23.

$$2.5\,kg \left(\frac{2.2\,lb}{1\,kg} \right)$$

22.

$$60 \left(\frac{15}{21} \right) \left(\frac{16}{13} \right)$$

24.

$$\left(\frac{45\,miles}{1\,hour} \right) \left(\frac{1\,hour}{60\,sec} \right) \left(\frac{1.6\,km}{1\,mile} \right)$$

25. 3.68 liter x 4.326 atm
26. (12.4 m) (1.2 m)
27. $(55 \times 10^4)(3.8 \times 10^2)$

28. $(0.043 \times 10^6)(0.02 \times 10^{-2})$
29. $(4.6 \times 10^{-8})(0.012 \times 10^{-3})$

For questions 30 to 34, rewrite the number in exponential form as indicated.

30. Rewrite 58.2×10^7 as a number times 10^5.
31. Rewrite 0.00356×10^{-4} as a number whose coefficient is between 1 and 10.
32. Rewrite 0.012×10^{-2} as a number times 10^{-4}.
33. Rewrite 46.267×10^{-3} as a number whose coefficient is between 1 and 10.
34. Rewrite 6.891×10^3 as a number times 10^6.

Division
The division of a number "a" by another number "b" can be expressed in several ways:

$$a \div b, \qquad \frac{a}{b} \qquad a/b$$

Fractions. Dividing a number by a fraction will give the same result as if the number were multiplied by the reciprocal, or inverse, of that fraction. The reciprocal of a fraction is formed by "flipping over" the fraction; for example, the reciprocal of 3/5 is 5/3 -. When numbers having units attached to them are divided, the units are expressed in the form of a "fraction" also.

Example

1.

$$15 \div \frac{5}{8} = 15 \; x \; \frac{8}{5} = \frac{15 \; x \; 8}{5} = \frac{120}{5} = 24$$

2.

$$\frac{24}{\frac{2}{5}} = 24 \; x \; \frac{5}{2} = \frac{24 \; x \; 5}{2} = \frac{120}{2} = 60$$

3.

$$\frac{75.6g}{14mL} = \frac{75.6}{14} \; x \; \frac{g}{mL} = 5.4\frac{g}{mL} = 5.4 \; g/mL$$

Numbers in Exponential Form. Reread Appendix 1 for a description of the procedure for dividing numbers expressed in exponential form.

Check Your Understanding

For questions 35 to 40, perform the indicated division.

35. $5.796 \div 1.68$

36. $\dfrac{32}{1.28}$

37. $24 \div \dfrac{6}{7}$

38. $\dfrac{14}{\frac{5}{8}}$

39. $\dfrac{478 \text{ miles}}{6.25 \text{ hr}}$

40. $\dfrac{0.275 \text{ mole}}{0.05 \text{ liter}}$

For questions 41 to 44, perform the indicated division and rewrite the answer as a number having a coefficient between 1 and 10.

41. $\dfrac{7196 \times 10^{12}}{2.57 \times 10^{25}}$

42. $\dfrac{125 \times 10^{8}}{3125 \times 10^{2}}$

43. $$\dfrac{37.5 \times 10^2}{0.025 \times 10^{-7}}$$

44. $$\dfrac{3.2 \times 10^2}{1600 \times 10^6}$$

Percent Notation

The term "percent" means "out of 100", and is denoted by the symbol "%". Thus, a statement such as "50% of the shirts are red" means that 50 out of 100 shirts (or half of the shirts) are red. This proportion can also be written 50:100 or 50/100. To convert numbers between percentage notation and the other types of numbers we have discussed, use the following rules:

1. To write a decimal number as a percentage, move the decimal point 2 places to the right and add a percent sign. For example, $0.48 = 48\%$, $0.0732 = 7.32\%$, and $2.781 = 278.1\%$.

2. To write a fraction as a percentage, first convert the fraction to a decimal number, by dividing the denominator into the numerator, and then convert the decimal to a percentage as described in Rule 1. For example,

$$\frac{3}{4} = 0.75 = 75\%$$

3. To write a percentage as a decimal number, move the decimal point 2 places to the left and remove the percent sign. For example, $71\% = 0.71$, $0.21\% = 0.0021$, and $130\% = 1.3$.

Check Your Understanding ───────────────────────────────

For questions 45 to 50, rewrite the given numbers as percents.

45. 0.46

47. 0.045

49. 2.671

46. 4/5

48. $\dfrac{2}{9}$

50. 3.18×10^{-1}

For questions 51 to 54, convert the percentages to decimals.

51. 4%

52. 20.5%

53. 7.67%

54. 0.03%

Solving Equations for an Unknown

An equation is a mathematical expression stating that two quantities are equal:

$$\frac{30}{36} = \frac{5}{6} \quad \text{or} \quad (12 \text{ eggs}) = (1 \text{ dozen eggs}).$$

Often we can state that two quantities are equal even when we don't know the exact value of some numbers in the equation. Such unknown values are represented by letters:

$$\frac{X}{4} = \frac{20}{8}, \quad \text{or} \quad P_1 V_1 = P_2 V_2$$

The process of figuring out the unknown value in an equation is called "solving the equation for the unknown." To solve an equation for an unknown quantity, there are only two rules to follow:

1. Try to undo what is being done to the unknown in the equation.

2. Always do the same thing to both sides of the equation, so that the two sides will remain equal.

Example

1. Solve the following equation for the value of Z: $\dfrac{Z}{4} = \dfrac{20}{8}$

In this equation, Z is being divided by 4. To undo this we will have to multiply by 4 ; however, we must remember to multiply both sides of the equation at the same time.

$$4 \times \frac{Z}{4} = 4 \times \frac{20}{8} \quad \text{or,} \quad Z = \frac{4 \times 20}{8} = 10$$

2. Solve the following equation for P_2: $P_1 V_1 = P_2 V_2$.

We see that P_2 is being multiplied by V_2. To undo this, we will have to divide by V_2, remembering to do the same thing to both sides of the equation.

$$\frac{P_1 V_1}{V_2} = \frac{P_2 V_2}{V_2} \quad \text{Or,} \quad \frac{P_1 V_1}{V_2} = P_2$$

11

For questions 55 to 58, solve the equation for X.

55. $\dfrac{X}{25} = \dfrac{760}{950}$

57. $\dfrac{275}{X} = \dfrac{5}{0.04}$

56. $16X = 86.4$

58. $\dfrac{X}{c+d} = \dfrac{ab}{Y}$

For questions 59 to 63, solve the equation for the indicated unknown.

59. $V_1M_1 = V_2M_2$; solve for M_2.
60. $K = °C + 273$; solve for $°C$.
61. $PV = nRT$; solve for V, and then solve for T.
62. $1.8(°C) = (°F) - 32$; solve for $°F$, and then solve for $°C$.
63. $\dfrac{P_1V_1}{T_1} = \dfrac{P_2V_2}{T_2}$; solve for P_2, then for V_2 , then for T_2.

Answers to Check Your Understanding Questions for Mathematics Review

1. 1.56×10^3 **2.** 7.432×10^6 **3.** 4.6×10^{-3} **4.** 1.45×10^2 **5.** 2.5×10^{-2} **6.** 7×10^{-7}
7. 4.56×10^5 **8.** 4.1×10^{-5} **9.** 2.5×10^4 **10.** 1.5×10^{-1} **11.** 31.05 **12.** 0.1045 **13.** 340.1×10^6
14. 186.2315 **15.** 405×10^{-5} **16.** 144.864 **17.** 6.475 **18.** 120×10^5 **19.** 60×10^{-6}
20. 1.5×10^{-8} **21.** $264/420 = 22/35$ **22.** $14{,}400/273 = 4800/91$ **23.** 5.5 lb **24.** 1.2 km/sec
25. 15.91968 liter-atm **26.** 14.88 m^2 **27.** 2.09×10^8 **28.** 8.6 **29.** 5.52×10^{-13} **30.** 5820×10^5
31. 3.56×10^{-7} **32.** 1.2×10^{-4} **33.** 4.6267×10^{-2} **34.** 0.006891×10^6 **35.** 3.45 **36.** 25 **37.** 28
38. 22.4 **39.** 76.48 miles/hr **40.** 5.5 moles/liter **41.** 2.8×10^{-10} **42.** 4×10^4 **43.** 1.5×10^{12}
44. 2×10^{-7} **45.** 46% **46.** 80% **47.** 4.5% **48.** 22% **49.** 267.1% **50.** 31.8% **51.** 0.04
52. 0.205 **53.** 0.0767 **54.** 0.0003 **55.** $X = 20$ **56.** $X = 5.4$ **57.** $X = 2.2$

58. $X = \dfrac{ab(c+d)}{Y}$ **59.** $M_2 = \dfrac{V_1M_1}{V_2}$ **60.** $°C = K - 273$

61. $V = \dfrac{nRT}{P}$; $T = \dfrac{PV}{nR}$ **62.** $°F = 1.8°C + 32$; $°C = \dfrac{°F - 32}{1.8}$

63. $P_2 = \dfrac{P_1V_1T_2}{T_1V_2}$; $V_2 = \dfrac{P_1V_1T_2}{T_1P_2}$; $T_2 = \dfrac{P_2V_2T_1}{P_1V_1}$

Chapter 1

THE STRUCTURE AND PROPERTIES OF MATTER

Chapters 1 through 5 introduce the basic chemical vocabulary used throughout the textbook. It is vital that you master the Important Terms within these chapters. If you have previously had a chemistry course, much of this will be review. You can check your knowledge by answering the self-test questions at the end of each section. If you have not had chemistry before, you will want to study these chapters slowly, becoming thoroughly familiar with each section before you go on. A little extra time spent studying these chapters now will make later chapters much easier to master.

1.1   What is Matter?

Matter has mass and occupies space. The amount of matter in an object determines the mass of the object. Mass is a measure of the resistance of an object to a change in its speed or in the direction of its motion. Weight is a measure of the gravitational attraction on an object. The mass of an object does not depend upon gravitational attraction, so it never changes.

Important Terms

matter   mass   weight

Check Your Understanding _____

For questions 1 to 3, answer true (T) or false (F).
1.   The mass of an object varies with a change in location.
2.   A Volkswagen has a greater mass than a Cadillac.
3.   If a stone weighs 20 pounds on the moon, it would weigh about 20 pounds on Earth.

1.2   Composition of Matter

Atoms are particles that make up matter. They are extremely small, and yet have been shown to be composed of many even smaller particles. The most important subatomic particles are the proton, the neutron, and the electron.

Important Terms

atom   proton   neutron   electron

1.3 Classes of Matter: Elements, Compounds, and Mixtures

Elements are the simplest substances encountered in the chemical laboratory. They make up all matter, either alone or in combination with other elements. Elements are pure substances that contain only one kind of atom. Compounds, by contrast, are pure substances composed of atoms of more than one element. A molecule is a particle composed of two or more atoms joined together.

Compounds, no matter how they are formed, always contain the same type of atoms in a definite ratio or proportion by weight. This is known as the Law of Definite Proportions. Mixtures, on the other hand, contain two or more substances in any proportion (such as salt in water or carbon monoxide in exhaust fumes). When substances combine to form a compound, a new substance with different properties is formed, but substances in a mixture keep their individual properties. Mixtures can be identified as either homogeneous or heterogeneous.

Important Terms

element	compound	atom
molecule	mixture	homogeneous
heterogeneous	Law of Definite Proportions	

Check Your Understanding _____

For questions 4 to 8, answer true (T) or false (F).

4. When heated in a laboratory, an element can break down to form two or more simpler substances.
5. A molecule is composed of two or more atoms.
6. Compounds contain atoms from two or more different elements.
7. Substances contained in a mixture can be identified by their individual properties.
8. An atom is the smallest particle of an element that keeps the properties of that element.
9. Identify each of the following as a homogeneous mixture (hom) or heterogeneous mixture (het).
   (a) concrete   (c) salad dressing   (e) honey   (g) sand
   (b) gasoline   (d) black coffee   (f) cytoplasm

1.4 Names and Symbols for the Elements

An element can be represented by its symbol, which often is the first letter of the element's name (for example, N for nitrogen and P for phosphorus). If two letters are used, the second letter is not capitalized (for example, Mg for magnesium and He for helium).

Check Your Understanding —————————————————————————————

Use the table on the inside back cover of the textbook to help you with questions 10 and 11.
10.   Write the symbol for each of the following elements.
       (a) hydrogen   (e) potassium   (h) copper
       (b) carbon   (f) manganese   (i) chlorine
       (c) oxygen   (g) iron   (j) zinc
       (d) sodium
11.   Write the names of each of the following elements.
       (a) N   (e) S   (h) Co
       (b) F   (f) Ca   (i) Se
       (c) Mg   (g) Cr   (j) Sn
       (d) P

1.5   The Three States of Matter

Matter exists in three states, solid, liquid, and gas, which can be converted one to another by adding or removing energy. When a physical change occurs, the nature of the substance remains the same--only its form is changed. In a chemical change, new substances with different chemical and physical characteristics are formed.

Important Terms

solid   liquid   gas
physical change   chemical change

Check Your Understanding —————————————————————————————

12.   Identify each of the following as a chemical change (C) or a physical change (P).
       (a)   Alcohol boiling
       (b)   Antacids neutralizing stomach acid
       (c)   Pulverizing an aspirin
       (d)   Dissolving an aspirin in water
       (e)   Wood burning
       (f)   Sugar being used by a cell to produce energy, carbon dioxide, and water
       (g)   Carbon monoxide reacting with oxygen to form carbon dioxide
       (h)   The sun drying swimmers on the beach

1.6   Scientific Method

The scientific method is a precise way of studying the world around us through observation, development of a hypothesis, testing that hypothesis, and then proposing a scientific theory.

Important Term

   scientific method

1.7   Accuracy and Precision

To be useful, data collected to test a hypothesis must be both accurate and precise. A precise measurement is one that is reproducible. An accurate measurement is one that comes close to the true value.

Important Terms

   accurate   precise

1.8   Significant Figures

Significant figures, or significant digits, indicate the precision with which a measurement was made. The correct number of significant figures to use in a measurement is the number of digits that are certain plus the first digit for which the measurement is uncertain.

Important Terms

   significant figures   significant digits

Check Your Understanding _____

13.   A certain volume of water (37.00 mL) was measured three times by each of three different students. Their recorded data was as follows:

Student I	40.00 mL	40.00 mL	40.00 mL
Student II	37.00 mL	38.20 mL	36.51 mL
Student III	37.05 mL	37.10 mL	36.95 mL

   (a)   Which set of measurements is the most precise: I, II, or III?
   (b)   Which set of measurements is the most accurate?
   (c)   How many significant figures are there in each of the measurements in III?

14.   Give the number of significant figures in each of the following measurements:
   (a) 225 lb   (c) 100 m   (e) 0.0036 mg
   (b) 22.50 g   (d) 0.207 liter   (f) 25.068 kg

1.9 Conversion Factor Method

Conversion factors, or unit factors, are ratios that are equivalent to the number "one." They are very important in mathematical calculations in this textbook. Conversion factors are used to change the units that are stated in a problem to the units that you want in the answer. A conversion factor can be formed from any equality by dividing both sides of the equation by one of the sides. For example, from the equality 1 in. = 2.54 cm, we can create two conversion factors:

$$\frac{1 \text{ in.}}{2.54 \text{ cm}} \quad \text{and} \quad \frac{2.54 \text{ cm}}{1 \text{ in.}}$$

Important Terms

      unit factor   conversion factor

Check Your Understanding

15.   Write two conversion factors for each of the following relationships.
    (a)   1 dozen eggs contains 12 eggs.
    (b)   1000 milliliters = 1 liter.
    (c)   There are 3 feet in a yard.
    (d)   1 centimeter = 0.39 inch.

1.10 The SI System of Units

The SI System of Units is used by scientists throughout the world. The SI system is identical to the metric system, but it uses different units for reference. For example, in the metric system the reference unit of volume is the liter and in the SI system it is the cubic meter. The SI and metric systems have an advantage over the English system in that all units of measure are related to their subparts by multiples of ten. The names of larger or smaller units are derived from the reference unit by using the appropriate prefix. The following are the basic units in the SI system, along with some other commonly used units.

Measurement	SI System	Common Units
Length	meter (m)	kilometer (km)
		centimeter (cm)
		millimeter (mm)
Mass	kilogram (kg)	gram (g)
		milligram (mg)
		microgram (μg)
Volume	cubic meter (m³)	liter (L)
		milliliter (mL)
		cubic centimeter (cc)

Most conversions between units in the SI system and units in the English system can be done using the conversion factors given in your text.

Important Terms

International System of Units
(SI system)

metric system
English system

Check Your Understanding ———————————————————

16. Complete the following table.

	Prefix	Multiple
(a)	kilo-	( )
(b)	centi-	( )
(c)	( )	10
(d)	milli-	( )
(e)	hecto-	( )
(f)	( )	0.1
(g)	( )	0.000001

17. For each of the following groups, place the units in order from the smallest to the largest.
    (a) centigram, microgram, kilogram, milligram
    (b) deciliter, milliliter, liter, centiliter
    (c) meter, micrometer, centimeter, millimeter

1.11 Length

The unit of length in the SI system is the meter, which is slightly longer than a yard (1 meter = 3.28 feet).

Important Terms

length

meter

Example ———————————————————

1. How many centimeters are there in 2.7 meters?

S The question in this problem is 2.7 meters = (?) centimeters.

T The relationship between centimeters and meters is:
    1 meter (m) = 100 centimeters (cm)

18

From this equation we can write two conversion factors:

$$\frac{1 \text{ m}}{100 \text{ cm}} \quad \text{and} \quad \frac{100 \text{ cm}}{1 \text{ m}}$$

Our problem asks: 2.7 meters = (?) centimeters; therefore, the second unit factor is the one to use.

E

$$2.7 \text{ m} \times \frac{100 \text{ cm}}{1 \text{ m}} = 270 \text{ cm}$$

P   Because the measurement "2.7 meters" has two significant figures, the answer must also have two significant figures. To clearly show that the answer has two significant figures, it should be written as the exponential number: 2.7×10^2.

2.   How many millimeters long is a 1.5 inch incision?

S   The question in this problem is 1.5 inch = (?) centimeters

T   The solution to this problem requires two relationships.

10 millimeters (mm) = 1 centimeter (cm)

Conversion factors: $\dfrac{10 \text{ mm}}{1 \text{ cm}}$ or $\dfrac{1 \text{ cm}}{10 \text{ mm}}$

and: 2.54 cm = 1 inch (in.)

Conversion factors: $\dfrac{2.54 \text{ cm}}{1 \text{ in}}$ or $\dfrac{1 \text{ in}}{2.54 \text{ cm}}$

The problem asks: 1.5 inch = (?) millimeters; therefore, we choose the conversion factors necessary to produce the answer in the correct units:

E

$$1.5 \text{ in.} \times \frac{2.54 \text{ cm}}{1 \text{ in.}} \times \frac{10 \text{ mm}}{1 \text{ cm}} = 38 \text{ mm}$$

P   Because the measurement "1.5 in." has two significant figures, the answer must also have two significant figures and is 38 mm.

Check Your Understanding

For questions 18 to 25, complete the conversions required.

18. 3.6 km	=	_____	m	22. 37.0 ft	=	_____	m
19. 125 cm	=	_____	m	23. 100 yd	=	_____	m
20. 0.250 m	=	_____	mm	24. 40 mi	=	_____	km
21. 468 m	=	_____	km	25. 65 mm	=	_____	in

1.12 Mass

The unit of mass in the SI system is the kilogram, which equals 2.2 pounds. The unit of mass in the metric system is the gram.

Important Terms

mass   gram   kilogram

Example

1. How many milligrams are there in a 0.18 gram sample of radioactive tracer?

S The problem asks: 0.18 g = (?) mg

T The relationship between grams and milligrams is:

1 gram (g) = 1000 milligrams (mg)

$$\text{Conversion factors:} \quad \frac{1\ \text{g}}{1000\ \text{mg}} \quad \text{or} \quad \frac{1000\ \text{mg}}{1\ \text{g}}$$

The problem asks: 0.18 g =(?) mg; therefore, the second conversion factor is the one to use.

E

$$0.18\ \cancel{\text{g}} \times \frac{1000\ \text{mg}}{1\ \cancel{\text{g}}} = 180\ \text{mg}$$

P The answer with two significant figures is 1.8×10^2 mg of radioactive tracer.

2. The weight limit for luggage on an international flight is 40 pounds, but the scale in the hotel is calibrated in kilograms. How many kilograms are there in 40 pounds?

S The problem asks 40 lb. = (?) Kg.

T   The relationship between pounds and kilograms is:

$$1 \text{ kilogram (kg)} = 2.2 \text{ pounds (lb)}$$

Conversion factors: $\dfrac{1 \text{ kg}}{2.2 \text{ lb}}$ or $\dfrac{2.2 \text{ lb}}{1 \text{ kg}}$

The problem asks: 40 lb = (?) kg; therefore, the first conversion factor is the one to use.

E
$$40 \text{ lb} \times \dfrac{1 \text{ kg}}{2.2 \text{ lb}} = 18 \text{ kg}$$

P   The answer should have two significant figures. It is 18 kg.

Check Your Understanding

For questions 26 to 35, complete the conversions required.

26.	0.4 g = _____ mg		31. 1.50 lb = _____ g
27.	2.4 kg = _____ g		32. 2.1 kg = _____ lb
28.	45 mg = _____ g		33. 30.0 oz = _____ kg
29.	8.7 g = _____ kg		34. 0.700 kg = _____ oz
30.	0.02 g = _____ µg		35. 6.80 oz = _____ g

1.13   Volume

The SI conversion of volume is the cubic meter, a fairly large conversion of measure. The liter (1000 liters = 1 cubic meter) is a more commonly used conversion of volume. One liter is slightly larger than a quart.

Important Terms

volume	liter	cubic meter
milliliter	cubic centimeter	

Example

1.   How many 5-milliliter doses are there in a bottle containing 0.3 liter of ampicillin?

S   In this problem we need to find the number of times 5 mL can be divided into 0.3 L

T   This problem can be solved in two steps. First, we answer the question 0.3 liter =(?) milliliters. Second, we must answer the question 300 mL =(?) doses. The relationship between liters and milliliters is.

1 liter = 1000 milliliters

Conversion factors: $\dfrac{1 \text{ liter}}{1000 \text{ mL}}$ and $\dfrac{1000 \text{ mL}}{1 \text{ liter}}$

The relationship between doses and milliliters is:

1 dose = 5 mL

Conversion factors: $\dfrac{1 \text{ dose}}{5 \text{ mL}}$ and $\dfrac{5 \text{ mL}}{1 \text{ dose}}$

E $0.3 \text{ L} \times \dfrac{1000 \text{ mL}}{1 \text{ L}} = 300 \text{ mL}$

$300 \text{ mL} \times \dfrac{1 \text{ dose}}{5 \text{ mL}} = 60 \text{ doses}$

P   The answer is 60 doses

2.   How many cups are there in 0.60 liter?

S   The question asks 0.60 L = (?) cups.

T   The two relationships needed to solve this problem are

1 liter = 1.06 quarts: $\dfrac{1 \text{ liter}}{1.06 \text{ qt}}$ or $\dfrac{1.06 \text{ qt}}{1 \text{ liter}}$

and

1 quart = 4 cups: $\dfrac{1 \text{ qt}}{4 \text{ cups}}$ or $\dfrac{4 \text{ cups}}{1 \text{ qt}}$

E   Using a conversion factor from each relationship, we can solve the problem in one step.

$0.60 \text{ L} \times \dfrac{1.06 \text{ qt}}{1 \text{ L}} \times \dfrac{4 \text{ cups}}{1 \text{ qt}} = 2.5 \text{ cups}$

P   The answer to two significant figures is 2.5 cups

For questions 36 to 45, complete the conversions required.

36.	250 dL = _____ liter	41. 9.0 gal = _____ liter
37.	150 mL = _____ cL	42. 2.50 pt = _____ mL
38.	2.2 liter = _____ mL	43. 405 mL = _____ qt
39.	425 mL = _____ liter	44. 250 mL = _____ cup
40.	3.4 cL = _____ liter	45. 0.50 cup = _____ liter

46. The doctor prescribes a dose of 1.5 teaspoons of penicillin three times a day until the bottle is empty. The bottle contains 250 mL of penicillin. For how many days will you be giving penicillin to your son? (1 teaspoon = 5.0 mL)

1.14   Temperature

There are three temperature scales used for measuring temperature. You are probably most familiar with the Fahrenheit scale. The Celsius scale is used in the metric system, and the Kelvin scale is used in the SI system. A one degree temperature difference is the same in the Celsius and Kelvin scales, but the scales start at different reference points. The freezing point of water is O°C and 273.15 K. Zero degrees on the Kelvin scale corresponds to absolute zero, the lowest temperature it is theoretically possible to reach.

Temperature conversion problems are slightly more complicated than conversions involving mass, volume or length. To convert a measurement from one system to another you must know two relationships. The first of these is the relative sizes of the units in the two systems. These are the conversion factors, such as 2.54 cm = 1 in., which you have learned. The relationship for temperature is 5 C° = 9 F°. The second relationship which you must know is what might be called a cross-over point. That is, some point at which you know what the measurement is in both systems. For temperature conversions we know that the freezing point of water is 32° F and 0° C. This is the cross-over point used in temperature conversions. For length, volume and mass, we know that zero in any system is zero in any other system. Therefore, we do not even need to include this in our conversion equation.

Important Terms

Fahrenheit scale   Kelvin scale   Celsius scale   absolute zero

Example ──

1. A child has a temperature of 104°F. What temperature is this on the Celsius scale?

S   The problem asks 104°F = (?)°C

T   To convert between the Fahrenheit and the Celsius scales, use the following relationship:

$$°C = \frac{5}{9} \, (°F - 32)$$

E   Substituting in this equation, we have

$$°C = \frac{5}{9} \, (104 - 32) = 40°C$$

P   The answer is 40°C

2.   Neon boils at 27 K. What temperature is this on the Celsius scale?

S   The question asks 27 K = (?)°C

T   To convert between the Kelvin and Celsius scales, use the following relationship: K = °C + 273 (actually 273.15, but for solving problems in this chapter we will use the conversion factor rounded to the nearest whole number).

E   Substituting in this relationship, we have

$$27 = °C + 273$$

$$°C = 27 - 273 = -246°C$$

P   The answer is -246°C. When adding numbers, the number of decimals in the answer equals the smallest number of decimals in the numbers added.

Check Your Understanding _____

47.   Do the following conversions:
   (a) 40°C = ____K   (c) 100 K = ____°C (e) 253 K = ____°F
   (b) 77°F = ____K   (d) 110°C = ____°F (f) 23°F = ____°C

1.17   Density and Specific Gravity

The density of a substance is the mass of that substance per unit of volume. The specific gravity of a substance compares the density of that substance to the density of water (density of water = 1 g/cc or 1 g/mL at 4°C).

$$density = \frac{mass}{Volume} \qquad specific\ gravity = \frac{density\ of\ sample}{density\ of\ water}$$

24

density   specific gravity

1.   A rock has a mass of 70 g. When it is placed in a 100 mL graduated cylinder containing water, the water level rises from the 41 mL mark to the 59 mL mark. What is the density of the rock?

S   We need to calculate the density of the rock in grams per cubic centimeter.

T   Density is determined by dividing the weight of the rock by its volume. First we must determine the volume of the rock and then we can determine its density. The volume of the rock is equal to the volume of the water displaced by the rock.

E

$$\text{volume} = 59 \text{ mL} - 41 \text{ mL} = 18 \text{ mL} = 18 \text{ cc}$$

$$\text{density of the rock} = \frac{70 \text{ g}}{18 \text{ cc}} = 3.88889 \text{ g/cc}$$

P   The calculator gives a result of 3.88889, but our answer must have only two significant figures. It is 3.9 g/cc

2.   An oil sample has a density of 0.82 g/cc. What is the specific gravity of the oil?

S   The problem asks us to convert a density of 0.82g/cc to specific gravity

T   Specific gravity equals the density of a solution divided by the density of water.

E

$$\text{specific gravity} = \frac{0.82 \text{ g/cc}}{1.00 \text{ g/cc}} = 0.82$$

P   The answer must have two significant figures and is 0.82. Remember that specific gravity does not have any units.

3.   Alcohol has a density of 0.79 g/cc. What is the weight of a 7.5 mL sample?

S   We must figure out the weight of the sample based on its density.

T   We can use the value of the density as a conversion factor to solve this problem.

$$\text{volume (mL)} \times \text{density (g/mL)} = \text{mass (g)}$$

E

$$7.5 \text{ mL} \times \frac{0.79 \text{ g}}{1 \text{ mL}} = 5.925 \text{ g (calculator answer)}$$

P   The answer to two significant figures is 5.9 g

Check Your Understanding

48. What is the density of a solid that has a volume of 23 mL and a mass of 163 g?
49. What is the specific gravity of a liquid that has a density of 0.950 g/cc? What is the mass of a 250-mL sample of this liquid?
50. A 75.0-mL sample of urine weighs 86.25 g. What is the specific gravity of this urine sample?
51. What is the volume of a solution that has a mass of 125 g and a density of 1.40 g/cc?
52. What is the density of a 25-g sample of solid if the water level rises from the 34 mL mark to the 47 mL mark when the sample is completely emersed in a graduated cylinder containing water?

Review Your Understanding

53. A box measures 4.00 inches on each side. What is its volume in cubic centimeters?
54. What is the length in centimeters of a 5.00 in. incision?
55. If a circular birthmark has a diameter of 0.25 inch, what is its radius in millimeters?
56. A patient is 5 feet 2 inches tall. What is her height in meters?
57. A sleeping pill contains 100 mg of powder. How many grams of powder is this?
58. What is the weight in kilograms of a 140-pound patient?
59. If the recommended dose of a drug is 4 milliliters, how many doses will there be in a bottle containing 0.2 liter?
60. How many grams of drug are there in 4 ampules if each ampule contains 7.5 grains (1 grain = 60 milligrams)?
61. A vial of Kanamycin is labeled 0.25 g/mL. The required dose of the drug is 15 mg for each kilogram the patient weighs. If a patient weighs 55 pounds, how many milliliters are necessary for the required dose?
62. If a physician ordered 2 grams of drug to be given in four equal doses and the drug were available in 250-mg capsules, how many capsules would you give for each of the four doses?

For questions 63 to 72 select the best answer.
63. A mass of 0.10 g can be expressed as:

(a)	100 mg	(c)	10.0 mg
(b)	1.00 mg	(d)	1000 mg

64. 0.232 lb is the same as:
   - (a) 210 g
   - (b) 105 g
   - (c) 0.470 g
   - (d) 470 g

65. If a person is 5 feet 4 inches tall, their height in centimeters is:
   - (a) 64 cm
   - (b) 25 cm
   - (c) 1630 cm
   - (d) 163 cm

66. 255 mL is the same volume as:
   - (a) 0.40 qt
   - (b) 0.270 qt
   - (c) 270 qt
   - (d) 241 qt

67. If a container holds 2.0 gallons, has what volume does it hold in the metric system?
   - (a) 8.0 L
   - (b) 7.5 L
   - (c) 6.5 L
   - (d) 8.8 L

68. A liquid has a density of 0.80 g/mL. What is the mass of 20.0 mL of this liquid?
   - (a) 0.04 g
   - (b) 16 g
   - (c) 25 g
   - (d) 0.04 g

69. If the density of a solution is 1.15 g/mL, what is the volume of 20.0 g of the solution?
   - (a) 21.2 mL
   - (b) 23.0 mL
   - (c) 17.4 g
   - (d) 20.0 mL

70. What is the density of a substance if 605 mL of this material has a mass of 225 g?
   - (a) 1.34 g/mL
   - (b) 0.372 g/mL
   - (c) 2.68 g/mL
   - (d) 1.50 g/mL

71. Convert -20°F to Celsius.
   - (a) -29°C
   - (b) -36°C
   - (c) -94°C
   - (d) -11°C

72. Convert -50°C to Fahrenheit.
   - (a) -28°F
   - (b) -122°F
   - (c) -90°F
   - (d) -58°F

73. Human body temperature is usually about:
   - (a) 99°C
   - (b) 37°F
   - (c) 37°C
   - (d) 72°F

74. On a summer day the temperature might reach 110°F in Las Vegas, what temperature is this on the Celsius scale?
   - (a) 61°C
   - (b) 93°C
   - (c) 140°C
   - (d) 43°C

75. What is the temperature in problem 72 on the kelvin scale?
   - (a) 223 K
   - (b) 323 K
   - (c) 215 K
   - (d) 183 K

Answers to Check Your Understanding Questions in Chapter 1

1. F 2. F 3. F 4. F 5. T 6. T 7. T 8. T 9.(a) het (b) hom (c) het (d) hom (e) hom (f) het (g) het 10.(a) H (b) C (c) O (d) Na (e) K (f) Mn (g) Fe (h) Cu (i) Cl (j) Zn 11.(a) nitrogen (b) fluorine (c) magnesium (d) phosphorus (e) sulfur (f) calcium (g) chromium (h) cobalt (i) selenium (j) tin 12.(a) P (b) C (c) P (d) P (e) C (f) C (g) C (h) P 13.(a) I (b) III (c) 4 14. (a) 3 (b) 4 (c) 1 (d) 3 (e) 2 (f) 5

15. (a) $\dfrac{1\ \text{dozen}}{12\ \text{eggs}}$, $\dfrac{12\ \text{eggs}}{1\ \text{dozen}}$ (b) $\dfrac{1000\ \text{mL}}{1\ \text{liter}}$, $\dfrac{1\ \text{liter}}{1000\ \text{mL}}$

(c) $\dfrac{3\ \text{ft}}{1\ \text{yd}}$, $\dfrac{1\ \text{yd}}{3\ \text{ft}}$ (d) $\dfrac{1\ \text{cm}}{0.39\ \text{in.}}$, $\dfrac{0.39\ \text{in.}}{1\ \text{cm}}$

16.(a) 1000 (b) 0.01 (c) deka- (d) 0.001 (e) 100 (f) deci- (g) micro 17. (a) micro- gram, milligram, centigram, kilogram (b) milliliter, centiliter, deciliter, liter (c) micrometer, millimeter, centimeter, meter 18. 3600 m 19. 1.25 m 20. 250 mm 21. 0.468 km (Watch the significant figures in your answers.) 22. 11.3 m 23. 91.5 m 24. 64 km 25. 2.6 in. 26. 400 mg 27. 2400 g 28. 0.045 g 29. 0.0087 kg 30. 20,000 g 31. 682 g 32. 4.6 lb 33. 0.851 kg 34. 24.6 oz 35. 193 g 36. 25 liter 37. 15 cL 38. 2200 mL 39. 0.425 liter 40. 0.034 liter 41. 34 liter 42. 1180 mL 43. 0.429 qt 44. 1.06 cup 45. 0.12 liter

46. $\dfrac{1.5\ \text{tsp}}{1\ \text{dose}} \times \dfrac{5.0\ \text{mL}}{1\ \text{tsp}} \times \dfrac{3\ \text{doses}}{1\ \text{day}} = \dfrac{22.5\ \text{mL}}{1\ \text{day}}$ $250\ \text{mL} \times \dfrac{1\ \text{day}}{22.5\ \text{mL}} = 11\ \text{days}$

47.(a) 313 K (b) 298 K (c) -173°C (d) 230°F (e) -4°F (f) -5°C 48. 7.1 g/cc 49. 0.950, 238 g 50. 1.15 51. 89.3 cc 52. 1.9 g/cc 53. 1050 cm^3 54. 12.7 cm 55. 3.2 mm 56. 1.6 m 57. 0.1 g 58. 63.6 kg 59. 50 doses

60. $7.5\ \text{grains} \times \dfrac{60\ \text{mg}}{1\ \text{grain}} \times \dfrac{1\ \text{g}}{1000\ \text{mg}} = 1.8\ \text{g}$

61. $55\ \text{lb} \times \dfrac{1\ \text{kg}}{2.2\ \text{lb}} \times \dfrac{15\ \text{mg}}{1\ \text{kg}} \times \dfrac{1\ \text{g}}{1000\ \text{mg}} \times \dfrac{1\ \text{mL}}{0.25\ \text{g}} = 1.5\ \text{mL}$

62. 2 capsules/dose

63. a, 64. b, 65. d, 66. b, 67. b, 68. b, 69. c, 70. b, 71. a, 72. d 73. c , 74. d, 75. a

Chapter 2   ENERGY

In this chapter we define energy and look at different types of energy and the characteristics of each type of energy. The material covered in this chapter will show up repeatedly in later chapters, as we study such topics as the states of matter, nuclear transformations and the reactions of living organisms.

2.1   What is Energy?

Energy is the ability or capacity to cause various kinds of change. For example, pedaling a bicycle up a hill requires your muscles to expend energy to move the bicycle from one place to another.

Important Term

> energy

2.2   Kinetic Energy

Kinetic energy is the energy an object possesses because of its motion. The amount of kinetic energy possessed by an object depends upon the mass of the object and the speed with which it is moving (its velocity). Because the particles that make up matter are in constant motion, they possess kinetic energy. The temperature of a substance is a measure of the average kinetic energy of all the particles that make up that substance. The higher the temperature of a substance, the greater the kinetic energy of its particles.

Important Terms

> kinetic energy   temperature

Check Your Understanding ─────────────────────────────

1.   Indicate which choice in each of the following pairs has greater kinetic energy.
   - (a)   A moving freight train or a freight train that is standing still.
   - (b)   A football player just before the ball is snapped or that same player just after the ball is snapped.
   - (c)   50 g of liquid water or 50 g of ice.

2.3   Potential Energy

Potential energy is stored energy. A substance can possess potential energy by virtue of its position. A truck parked at the top of a hill possesses potential energy that would be converted to kinetic energy if its brakes were to fail, causing it to roll down the hill. A hibernating animal contains potential energy in its stored body fat; this energy is slowly released to sustain the animal over the winter.

Important Term

>   potential energy

Check Your Understanding ————————————————————————

2. Indicate which choice in each of the following pairs has greater potential energy:
    (a) An arrow in a drawn bow or the arrow after it leaves the bow.
    (b) A football player just before the ball is snapped or the same player just after the ball is snapped.
    (c) A person at the top of a flight of stairs or that same person at the bottom of the stairs.
3. List the following in order of increasing kinetic energy:
    (a) water at 25°C   (c) water at 325 K   (e) water at 240°F
    (b) water at 28°F   (d) water at -75°C   (f) water at 140 K

2.4 Heat Energy

Heat is energy that is transferred from one place to another because of a difference in temperature. Heat is measured in units called calories. A calorie is the amount of heat energy required to raise the temperature of one gram of water one degree Celsius. In the SI system, the unit of heat energy is the joule (1 cal = 4.184 J).

The specific heat of a substance is the amount of energy required to raise the temperature of one gram of a substance one degree Celsius.

$$\text{specific heat} = \frac{\text{calories}}{\text{g } °C}$$

The amount of heat necessary to change the temperature of a sample of a substance depends upon the specific heat of the substance, the mass of the sample, and the change in the temperature.

$$\text{calories} = g \times \Delta t \times \frac{\text{cal}}{\text{g } °C}$$

A food Calorie is equivalent to 1000 calories, or one kilocalorie. Notice that the food Calorie is capitalized to distinguish it from the calorie. The number of kilocalories of energy in a sample of food can be measured in a calorimeter. The basal metabolism rate is the minimum amount of energy required daily to maintain a human body at rest.

Important Terms

heat   calorie   kilocalorie
joule   specific heat   calorimeter
basal metabolism rate (BMR)

Example _____

When a 6-gram sample of butter is burned in a calorimeter, the temperature of the 500 mL of water in the calorimeter increases from 27°C to 63°C. How many kilocalories are contained in one gram of butter? (Remember that the density of water = 1 g/mL.)

S   The question asks us to determine the number of kilocalories in one gram of the butter sample.

T   To solve this problem we use the equation,

$$\text{cal} = \text{g of water} \times \Delta t \times \frac{1 \text{ cal}}{\text{g } °C}$$

Use of this equation tells us how many calories there are in the 6 g sample. We then divide by 1000 to determine kilocalories and by 6 to determine the number of kilocalories in 1 g of butter.

E

$$\text{cal} = 500 \text{ g} \times (63°C - 27°C) \times \frac{1 \text{ cal}}{\text{g } °C}$$

$$= \frac{500 \cancel{\text{g}} \times 36°\cancel{C} \times 1 \text{ cal}}{\cancel{\text{g}} \ °\cancel{C}}$$

$$= 18{,}000 \text{ cal or } 18 \text{ kcal}$$

P   If 6 grams of butter released 18 kilocalories of energy then one gram of butter would contain 18/6 or 3 kilocalories.

Check Your Understanding _____

4.   Complete the following conversions:
   (a) 250 cal   = _____ kcal   (d) 9500 joule = _____ kcal
   (b) 3.4 kcal   = _____ cal   (e) 325 cal   = _____ joule
   (c) 25 Calories = _____ kcal

5.   How many kilocalories are contained in one gram of hamburger if the energy released by a 10-gram sample of hamburger raised the temperature of 700 grams of water in a calorimeter from 24 to 64°C?

6.   The oxidation of one gram of carbohydrate produces 4 kcal. The oxidation of 6 grams of carbohydrate will raise the temperature of 800 grams of water how many degrees?

7.   The oxidation of one gram of fat produces 9 kcal. If 8 grams of fat are oxidized in a calorimeter containing 1500 grams of water, what will be the final temperature of the water if the initial temperature is 30°C?

8.   The oxidation of a gram of protein produces 4 kcal. How many grams of protein must be oxidized to raise the temperature of 1300 grams of water in a calorimeter from 23°C to 29°C?

9.   How many kilocalories of heat are produced during the burning of a sample of food if the temperature of 1400 grams of water in a calorimeter is raised from 32°C to 38°C?

2.5   Changes in State

Matter exists in three states: solid, liquid, and gas. A change of state occurs when matter goes from one state to another. Changes in state can require energy as when ice melts or can give off energy as when gas condenses. Changes that require energy are called endothermic and those that give off energy are called exothermic.

Important Terms

endothermic   exothermic

2.6   Electromagnetic Energy

The visible light that we see is only a tiny part of the entire range of energy called electromagnetic energy or electromagnetic radiation. This energy travels in waves. Electromagnetic radiation with long wavelengths (for example, radio waves) has low energy, but electromagnetic radiation with short wavelengths (for example, gamma rays) has extremely high energy. Electromagnetic radiation has different uses and different effects on living organisms depending upon its energy.

Important Terms

electromagnetic spectrum   wavelength
microwave radiation   visible light
ultraviolet radiation   gamma rays
infrared radiation   X rays

For questions 10 to 18, choose the best answer.

10. Artificial high energy radiation is
   (a) gamma rays
   (b) X rays
   (c) radio waves
   (d) infrared radiation

11. The radiation with the lowest energy is
   (a) radio waves
   (b) gamma rays
   (c) microwave radiation
   (d) infrared radiation

12. This radiation causes sunburn.
   (a) gamma rays
   (b) red light
   (c) ultraviolet radiation
   (d) infrared radiation

13. This high energy radiation is produced by natural sources.
   (a) radio waves
   (b) gamma rays
   (c) microwave radiation
   (d) X rays

14. The radiation with the longest wavelength is
   (a) radio waves
   (b) gamma rays
   (c) microwave radiation
   (d) X rays

15. This visible light has the highest energy.
   (a) red light
   (b) blue light
   (c) yellow light
   (d) purple light

16. Highly penetrating radiation is
   (a) radio waves
   (b) gamma rays
   (c) microwave radiation
   (d) infrared radiation

17. This radiation is given off by warm objects
   (a) X rays
   (b) radio waves
   (c) ultraviolet radiation
   (d) infrared radiation

18. This electromagnetic radiation is used for cooking.
   (a) gamma rays
   (b) X rays
   (c) microwave radiation
   (d) infrared radiation

2.7   Law of Conservation of Energy

In any process, energy must be conserved. This means that the total amount of energy at the end of a process must equal the total amount of energy at the beginning of the process. Energy can neither be created nor destroyed; it can only change form. Each of these statements is a way of saying the Law of Conservation of Energy, or the First Law of Thermodynamics.

For example, the potential energy stored in a log does not disappear when the log is burned, but rather, it is converted into heat and light energy. Most of the light energy that reaches the earth from the sun is converted into heat energy, but a small amount is absorbed by green plants and converted into chemical energy.

Important Terms

        Law of Conservation of Energy   First Law of Thermodynamics

2.8   Entropy

All naturally occurring processes tend toward a state of lower energy and higher disorder or randomness. The term given to this disorder or randomness is entropy. The more disordered a system, the greater is its entropy. The Second Law of Thermodynamics states that the entropy or disorder of the universe is increasing.

In order for nonspontaneous reactions (those reactions that go toward higher energy and lower entropy) to occur, they must be coupled with reactions that produce energy. For every system of such coupled reactions, we find that the total entropy of the system increases.

A living organism contains highly complex and organized structures and is, therefore, in a state of very low entropy. The natural tendency is for these structures to break down, so that the living organism must constantly expend energy from food that it consumes to maintain its state of low entropy. When the organism dies and can no longer generate this energy, the natural process toward increased entropy (that is, decay) begins.

Important Terms

      entropy   spontaneous process
      Second Law of Thermodynamics   nonspontaneous process

Check Your Understanding ────────────────────────────────

For questions 19 to 22, fill in the blank with the correct word or words.
19.   The word that describes the disorder of a system is _____.
20.   All naturally occurring processes tend toward a state of _____ energy and _____ entropy.
21.   When a process occurs, the energy at the end of the process must _____ the energy at the beginning of the process.
22.   The randomness of the universe is _____.
23.   Which one in each of the following pairs has the higher entropy?
        (a)   A new deck of cards, or a shuffled deck of cards
        (b)   Water as a liquid, or water as a gas

    (c)   A sugar cube, or sugar dissolved in a cup of coffee

    (d)   Compost, or the grass in a lawn

Answers to Check Your Understanding Questions in Chapter 2

1. (a) a moving freight train (b) after the ball is snapped   (c) 50 g of liquid water

2. (a) arrow in the drawn bow (b) before the ball is snapped (c) at the top of the steps

3. f, d, b, a, c, e **4.** (a) 0.250 kcal (b) 3400 cal (c) 25 kcal (d) 2.3 kcal (e) 1360 cal

5. 2.8 kcal

$$700 \text{ g} \times (64 \text{°C} - 24 \text{°C}) \times \frac{1 \text{ cal}}{\text{g °C}} = 28{,}000 \text{ cal} = 28 \text{ kcal}$$

28 kcal/10 g = 2.8 kcal/1 g

6. 30°C

$$6 \text{ g} \times \frac{4 \text{ kcal}}{\text{g}} \times \frac{1000 \text{ cal}}{1 \text{ kcal}} \times \frac{1 \text{ g °C}}{\text{cal}} \times \frac{1}{800 \text{ g}} = 30 \text{°C}$$

7. 78°C

$$8 \text{ g} \times \frac{9 \text{ kcal}}{\text{g}} \times \frac{1000 \text{ cal}}{1 \text{ kcal}} \times \frac{1 \text{ g °C}}{\text{cal}} \times \frac{1}{1500 \text{ g}} = 48 \text{°C temperature change}$$

initial temperature = 30°C

final temperature = 30°C + 48°C = 78°C

8. 2.0 g protein

$$1300 \text{ g} \times (29 \text{°C} - 23 \text{°C}) \times \frac{1 \text{ cal}}{\text{g °C}} \times \frac{1 \text{ kcal}}{1000 \text{ cal}} = 7.8 \text{ kcal}$$

$$7.8 \text{ kcal} \times \frac{1 \text{ g protein}}{4 \text{ kcal}} = 1.95 \text{ g} \quad (2.0 \text{ g to two significant figures})$$

9. 8.4 kcal

$$1400 \text{ g} \times \frac{1 \text{ cal}}{\text{g °C}} \times (38 \text{°C} - 32 \text{°C}) \times \frac{1 \text{ kcal}}{1000 \text{ cal}} = 8.4 \text{ kcal}$$

10. b **11.** a **12.** c **13.** b **14.** a **15.** d **16.** b **17.** d **18.** c **19.** entropy **20.** lower, higher **21.** equal **22.** increasing **23.** (a) shuffled cards (b) water as a gas (c) dissolved sugar (d) compost

Chapter 3   THE THREE STATES OF MATTER

In this chapter, we discuss the three states in which matter can exist: solid, liquid, and gas. We examine how they differ and how they can be converted one to another. We also discuss the laws that describe the behavior of matter in the gaseous state.

3.1   Solids

A solid has a definite shape and a definite volume. Its particles are close together and arranged either in a disordered fashion (an amorphous solid), or in a highly ordered fashion (a crystalline solid.) Solid particles have restricted motion and can only vibrate around a fixed point. As a solid is heated, the particles vibrate more and more violently. Finally, the temperature is reached at which the structure of the solid begins to break down. This temperature is the melting point of the solid. The heat of fusion is the amount of energy required to change one gram of a solid into a liquid at the melting point.

Important Terms

solid   amorphous solid   crystalline solid
melting point   heat of fusion

Example _____

How many kilocalories of energy are required to melt 150 grams of benzene if the heat of fusion of benzene is 30 cal/g?

S   The problem asks you to determine the amount of heat, in kilocalories, required to melt the sample.

T   To solve the problem use the heat of fusion for benzene to determine the amount of heat required, then convert the answer, in calories, into kilocalories:

$$\text{heat of fusion of benzene} = 30 \text{ cal/g}$$
$$1000 \text{ cal} = 1 \text{ kilocalorie}$$

E

$$150 \text{ \cancel{g}} \times \frac{30 \text{ \cancel{cal}}}{1 \text{ \cancel{g}}} \times \frac{1 \text{ kcal}}{1000 \text{ \cancel{cal}}} = 4.5 \text{ kcal}$$

P   The answer has two significant figures and is 4.5 kilocalories

3.2   Liquids

A liquid has a definite volume but not a definite shape; it will take on the shape of the container. Particles in a liquid are close together, but are not held in place as tightly as in a solid and can slide over one another. Two properties of a liquid resulting from the attraction between the liquid particles are viscosity (a measure of how easily a liquid flows) and surface tension (the resistance of the particles on the surface of a liquid to the expansion of that liquid). Very energetic particles near the surface of a liquid can escape to the gaseous state in a process called evaporation. The opposite of evaporation is condensation: the formation of liquid as gas particles are cooled. Sublimation is a change of state in which a solid goes directly into the gaseous state.

As heat is added to a liquid, the particles move more and more rapidly until a temperature is reached at which bubbles of vapor are formed within the liquid. Under conditions of normal atmospheric pressure, this temperature is known as the normal boiling point of the liquid. The heat of vaporization of a liquid is the amount of energy required to change one gram of a liquid into a gas at the boiling point of the liquid.

Important Terms

liquid	viscosity	surface tension
evaporation	condensation	boiling point
sublimation	heat of vaporization	boiling

Example _____

1.   How many kilocalories are released when 33 grams of ethanol are condensed (converted from a gas to a liquid) if the heat of vaporization of ethanol is 200 cal/g?

S   This problem is solved in the same way as the previous example, except that it involves the condensation of a vapor.

T   Use the heat of fusion of ethanol to determine the number of calories released and then convert that answer into kilocalories.

$$\text{heat of fusion of ethanol} = 200 \text{ cal/g}$$
$$1000 \text{ calories} = 1 \text{ kcal}$$

E

$$33 \cancel{g} \times \frac{200 \cancel{cal}}{1 \cancel{g}} \times \frac{1 \text{ kcal}}{1000 \cancel{cal}} = 6.6 \text{ kcal}$$

P   The answer has two significant figures and is 6.6 kcal

2. How many kilocalories are released when 25.0 g of steam at 100° C is condensed and then allowed to cool to 80° C?

S   The problem asks for the amount of heat in kilocalories released when the sample condenses and then cools.

T   This problem has three parts. You must determine the amount of heat released due to condensation, the amount released as the sample cools from 100° C to 80° C, and then the answer must be converted into kilocalories.

$$\text{heat of condensation of water} = 540 \text{ cal/g}$$
$$\text{specific heat of water} = 1 \text{ cal/g-°C}$$

E $\text{heat released} = (25.0 \text{ g} \times 540 \dfrac{\text{cal}}{\text{g}}) + (25.0 \text{ g} \times 20°C \times \dfrac{1 \text{ cal}}{\text{g-°C}})$

heat released = 13,500 cal + 500 cal = 14,000 cal

P   heat released = 14.0 kcal

Check Your Understanding _____

For questions 1-12, indicate whether the statement is true (T) or false (F).

1. Particles of a substance in the solid state have more kinetic energy than particles of that substance in the liquid state.
2. Particles in the gaseous state are much more widely separated than particles in the liquid state.
3. At normal atmospheric pressure, the temperature of boiling water will remain at 100°C no matter how much heat energy is added to the water.
4. A liquid will take on the shape of the container, because liquid particles are held in a fixed position.
5. A diamond is an example of an amorphous solid.
6. Oil is less viscous than water.
7. The higher the surface tension of a liquid, the more the particles on the surface of the liquid will prevent the liquid from spreading out.
8. Adding turpentine to some paints will make the paint less viscous.
9. The temperature at which a liquid becomes a gas is its melting point.
10. When particles of a liquid evaporate, the average kinetic energy of the liquid increases.
11. Lowering the temperature of a gas increases the kinetic energy of the gas particles.

12. A real gas does not behave like an ideal gas when the temperature of the gas is very low.

13. Indicate whether each of the following processes are endothermic or exothermic.
(a)	boiling	(d)	condensation
(b)	sublimation	(e)	melting
(c)	freezing	(f)	evaporation

14. How much heat energy is required to convert 1.00 L of water at 100°C to steam at 100°C? (Remember that the density of water = 1 g/mL.)

3.3 Gases

The pressure exerted by a gas results from the collisions of the gas particles against the sides of the container. The definition of pressure is force per unit area. Pressure can be measured in many units. You should know the definition of an atmosphere, millimeters of mercury, torr, and pascal.

Important Terms

atmosphere (atm)   millimeters of mercury (mm Hg)
torr (torr)   standard atmospheric pressure
pascal (Pa)   standard temperature and pressure (STP)

Check Your Understanding ———————————————————————————

For questions 15 to 20, complete the conversions required.

15. 327 mm Hg = _____ atm   18. 1102 torr = _____ atm

16. 157 torr   = _____ mm Hg   19. 2.7 atm = _____ mm Hg

17. 0.850 atm = _____ torr   20. 56 torr = _____ Pa

3.4 Boyle's Law: The Relationship Between Pressure and Volume

Boyle's law describes the relationship between the volume and the pressure of a gas. If the pressure on a gas is increased, the gas will be squeezed into a smaller volume. Mathematically, Boyle's law states that at a constant temperature, pressure times volume equals a constant (PV = constant). To solve problems using Boyle's law, we make use of the following equation:

$$P_1V_1 = P_2V_2 \quad \text{(at constant T)}$$

Important Term

Boyle's law

1. If a gas occupies a volume of 755 mL at standard atmospheric pressure, what will be the volume of the gas if the pressure is increased to 955 mm Hg and the temperature remains constant?

S   In this problem we are asked to find the value of V_2: $V_2 = (?)$

T   We must first identify P_1, P_2, V_1, and V_2. (Note: Be certain the two pressures are expressed in the same units, and the two volumes are expressed in the same units.)

$$P_1 = 760 \text{ mm Hg} \qquad P_2 = 955 \text{ mm Hg}$$
$$V_1 = 755 \text{ mL} \qquad V_2 = ?$$

Next ask yourself: will the starting volume of 755 mL increase or decrease if the pressure is increased from 760 mm Hg to 955 mm Hg? Boyle's law states that the volume will decrease if the pressure is increased. Therefore, we must multiply the volume by a ratio of pressures that will decrease the volume.

E

$$755 \text{ mL} \times \frac{760 \text{ mm Hg}}{955 \text{ mm Hg}} = 600.8377 \text{ mL (calculator)}$$

P   The answer to three significant figures is 601 mL, which meets our prediction of a decrease in volume.

Note: You can also solve this problem by using the equation for Boyle's law.

$$760 \text{ mm Hg} \times 755 \text{ mL} = 955 \text{ mm Hg} \times V_2$$

$$\frac{760 \text{ mm Hg} \times 755 \text{ mL}}{955 \text{ mm Hg}} = V_2$$

$$601 \text{ mL} = V_2$$

2. A 25-liter sample of carbon dioxide that exerts a pressure of 755 torr has been compressed into a 5.0-liter cylinder. If the temperature remains constant, what is the pressure of the carbon dioxide in the cylinder?

S   The problem asks for the final pressure: $P_2 = ?$

T   This problem is solved following the same procedure as example 1.

$$P_1 = 755 \text{ torr} \qquad P_2 = ?$$
$$V_1 = 25 \text{ liters} \qquad V_2 = 5.0 \text{ liters}$$

If the volume of the gas decreases, the pressure will increase.

E

$$755 \text{ torr} \times \frac{25 \text{ liters}}{5.0 \text{ liters}} = 3775 \text{ torr}$$

Or, by use of the Boyle's law equation:

$$755 \text{ torr} \times 25 \text{ liters} = P_2 \times 5.0 \text{ liters}$$

$$\frac{760 \text{ torr} \times 25 \text{ liters}}{5.0 \text{ liters}} = P_2$$

$$3775 \text{ torr} = P_2$$

P   The answer should have only two significant figures and is 3800 torr. (3.8×10^3 torr)

Check Your Understanding —————————————————————————

21.   A weather balloon is inflated to a volume of 2.00×10^2 L on a day when the atmospheric pressure is 755 mm Hg. What would be the volume of the balloon at an altitude of 17,000 feet where the atmospheric pressure is 370 mm Hg? (Assume that the temperature remains constant.)
22.   A sample of oxygen is placed in a 600-mL cylinder at a pressure of 1.0 atm and the volume of the cylinder is then compressed to 80 mL. What is the new pressure in the cylinder if the temperature remains constant?
23.   A gas occupies a volume of 280 mL at a pressure of 450 torr. If the gas is transferred to a 1.5 L container, what is the new pressure (in atm) exerted by the gas?

3.5   Charles's Law: The Relationship Between Volume and Temperature
Charles's law describes the relationship between the volume of a gas and the temperature of the gas (in degrees Kelvin) when the pressure remains constant. If the temperature of a gas is increased, the volume of the gas will increase.

Mathematically, we can express the relationship as follows:

$$\frac{V_1}{T_1} = \frac{V_2}{T_2} \quad \text{(at constant P)}$$

Important Term

Charles's law

Example _____

1. A balloon contains 3.5 liters of helium at 30°C. What would be the volume of the helium if it is cooled to 10°C without changing the pressure?

S The question asks for the value of P_2: $P_2 = (?)$

T First we must identify V_1, T_1, V_2, and T_2. (Remember that all temperatures must be expressed in Kelvins.)

$$V_1 = 3.5 \text{ liters} \qquad V_2 = \text{ ?}$$
$$T_1 = 303 \text{ K } (30°C + 273) \qquad T_2 = 283 \text{ K}$$

Next ask yourself: will the volume increase or decrease if the temperature decreases from 30°C to 10°C? From Charles's law we know that the volume will decrease when the temperature decreases. So we must multiply the volume by a ratio of temperatures that will decrease the volume.

E

$$3.5 \text{ L} \times \frac{283 \text{ K}}{303 \text{ K}} = 3.3 \text{ L}$$

P The answer of 3.3 L has the correct number of significant figures.

This problem could also be solved by substituting the values into the equation for Charles's law.

$$\frac{3.5 \text{ L}}{303 \text{ K}} = \frac{V_2}{283 \text{ K}}$$

$$3.3 \text{ L} = V_2$$

2. To what Celsius temperature must a gas be heated to triple its volume if it occupies 452 mL at 30°C? (The pressure is constant.)

S We need to find the correct value for T_2: $T_2 = $ (?)

T Following the same steps as example 1:

$$V_1 = 452 \text{ mL} \qquad\qquad V_2 = 1356 \text{ mL}$$
$$T_1 = 303 \text{ K} \qquad\qquad T_2 = \text{?}$$

For the volume of a gas to increase while the pressure remains constant, the temperature must increase.

E

$$303 \text{ K} \times \frac{1356 \text{ mL}}{452 \text{ mL}} = 909 \text{ K}$$

This problem could also be solved by substituting the values into Charles's law:

$$\frac{452 \text{ mL}}{303 \text{ K}} = \frac{1356 \text{ mL}}{T_2}$$

$$452 \text{ mL} \times T_2 = 1356 \text{ mL} \times 303 \text{ K}$$

$$T_2 = \frac{1356 \text{ mL} \times 303 \text{ K}}{452 \text{ mL}}$$

$$T_2 = 909 \text{ K}$$

P The answer is 909 K or 636° C

3. An 11 L sample of neon at 1.0 atm and -198°C expands to 22 L at -173°C. What is the new pressure?

S The problem asks for the value of P_2: $P_2 = $ (?)atm

T (a) To solve this problem we identify all the variables:

$$P_1 = 1.0 \text{ atm} \qquad\qquad P_2 = \text{?}$$
$$V_1 = 11 \text{ L} \qquad\qquad V_2 = 22 \text{ L}$$
$$T_1 = 75 \text{ K} \qquad\qquad T_2 = 100 \text{ K}$$

(b) We can solve this problem by reasoning in two steps. First, the volume is increasing; from Boyle's law we know this should make the pressure decrease. Therefore, we multiply P_1 by a ratio of volumes that will decrease the pressure. Second, the temperature is increasing. The higher the temperature, the greater the kinetic energy of the gas particles. Therefore, the greater the pressure. So the ratio of temperatures to use is the one that will increase the pressure.

E

$$1.0\text{atm} \times \frac{11 \cancel{L}}{22 \cancel{L}} \times \frac{100 \cancel{K}}{75 \cancel{K}} = 0.67 \text{ atm}$$

P   0.67 has two significant figures and is the correct answer.

This problem could also be solved by substituting in the combined gas law equation:

$$\frac{1.0 \text{ atm} \times 11 \text{ liters}}{75 \text{ K}} = \frac{P_2 \times 22 \text{ liters}}{100 \text{ K}}$$

$$\frac{1.0 \text{ atm} \times 11 \cancel{\text{liters}} \times 100 \cancel{K}}{75 \cancel{K} \times 22 \cancel{\text{liters}}} = P_2$$

$$0.67 \text{ atm} = P_2$$

Check Your Understanding _____

24.   A sample of helium gas occupies a volume of 250 mL at 37°C. What will be the new Celsius temperature of the helium if it is condensed to one tenth its original volume? (The pressure remains constant.)
25.   A 2.0 L sample of nitrogen gas is heated from 20°C to 90°C. What is the final volume of the nitrogen if the pressure remains constant?

3.6   Henry's Law
Henry's law relates the pressure of a gas to its solubility in a liquid. This law states that the higher the pressure on a gas, the greater its solubility in a liquid (when the temperature remains constant).

Important Terms

          Henry's law   the bends

3.7   Dalton's Law of Partial Pressures
Dalton's law states that the pressure exerted by a mixture of gases in a container is equal to the sum of the partial pressures of those gases. Each gas in the mixture acts independently of the others and will exert a pressure (called its partial pressure) as if it were alone in the

container. The pressure exerted by a gas is directly related to the number of gas particles in the container. Gas collected over water will contain a mixture of gas and water molecules. The pressure of the water molecules is called the vapor pressure, and will be dependent upon the temperature of the water.

Important Terms

Dalton's law   partial pressure   vapor pressure

Example _____

1.   You have a 515 mL container of helium at a pressure of 455 mm Hg, and a 515 mL container of oxygen at a pressure of 245 mm Hg. You now place both samples in the same 500 mL container. What is the partial pressure of each gas, and what is the total pressure in the container?

S   The problem asks for the partial pressure of oxygen (P_{oxygen}), the partial pressure of Helium (P_{Helium}), and the total pressure in the container (P_{total}).

$$P_{oxygen} = (?) \qquad P_{helium} = (?) \qquad P_{total} = (?)$$

T   Based on the definition of partial pressure, P_{helium} = 455 mm Hg and P_{oxygen} = 245 mm Hg.

Since the total pressure equals the sum of the partial pressures,

$$P_{total} = P_{helium} + P_{oxygen}$$

E $$P_{total} = 455 \text{ mm Hg} + 245 \text{ mm Hg} = 700 \text{ mm Hg}$$

P   The total pressure, P_{total}, should be written 7.00×10^2 to show that it has three significant figures.

2.   You collect a sample of hydrogen gas in a laboratory apparatus similar to Figure 3.10 in your text. The water temperature is 20°C and the barometer reads 738.5 mm Hg. The volume of the collected gas is 350 mL. Calculate the partial pressure of the hydrogen.

S   The question asks us to determine the partial pressure of hydrogen. $P_{hydrogen} = (?)$

45

T   Using Table 3.1 in your text, we see that the vapor pressure of water at 20°C is 17.5 torr. We know that 1 mm Hg = 1 torr, so the atmospheric pressure of 738.5 mm Hg = 738.5 torr. The volume of gas collected does not affect the total pressure nor the partial pressures of hydrogen and water.

$$P_{oxygen} = P_{total} - P_{water\ vapor}$$

E $P_{oxygen} = 738.5\ torr - 17.5\ torr = 721.0\ torr$

P   The answer 721.0 has the correct number of significant figures.

3.8   The Diffusion of Respiratory Gases

The diffusion of respiratory gases in our bodies is directly related to the partial pressures of those gases. The gases (carbon dioxide and oxygen) will diffuse from a region of higher partial pressure to a region of lower partial pressure.

3.9   The Kinetic Theory of Gases

The particles of a gas move very rapidly in a random chaotic fashion. A gas has neither a definite volume nor a definite shape, but will take on the volume and shape of a container. The kinetic theory, describing the properties of an ideal gas, can be summarized as follows:

1. The very small particles of a gas are widely separated from one another.
2. These particles are moving very fast in a random fashion and travel in straight lines until they collide with one another or with the sides of the container.
3. A particle of gas doesn't lose any energy when it collides with another particle or with the sides of the container.
4. Gas particles are not attracted or repulsed by other gas particles.
5. The kinetic energy of the gas particles increases with an increase in temperature.

Real gases behave according to this theory except under conditions of very low temperature or very high pressure.

Important Terms

      ideal gas   kinetic theory of gases   absolute zero

Check Your Understanding _____

For questions 26 to 33, choose the correct answer.

26. Pressure equals
    (a)   volume X temperature   (c)   force per unit area
    (b)   volume/temperature   (d)   temperature/volume
27. One atmosphere equals
    (a)   17.5 torr   (c)   1 mm Hg
    (b)   760 torr   (d)   133.3 Pa
28. Standard atmospheric pressure equals
    (a)   760 mm Hg   (c)   1 torr
    (b)   760 Pa   (d)   760 atm
29. If you decrease the volume of a gas when the temperature is constant, the pressure will
    (a)   increase   (c)   remain the same
    (b)   decrease   (d)   none of the above
30. If you increase the temperature of a gas when the pressure is constant, the volume will
    (a)   increase   (c)   remain the same
    (b)   decrease   (d)   none of the above
31. If the pressure of a gas above a liquid is increased, its solubility in the liquid will
    (a)   increase   (c)   remain the same
    (b)   decrease   (d)   none of the above
32. The total pressure of a mixture of gases will equal
    (a)   the atmospheric pressure minus the vapor pressure
    (b)   the sum of the partial pressures of the gases
    (c)   the sum of the partial pressures of the gases/volume
    (d)   the volume time the partial pressures
33. Absolute zero equals
    (a)   0°C   (c)   0 K
    (b)   -273 K   (d)   273°C

34. If a gas occupies 115 mL at a pressure of 763 torr, what volume will it occupy if the pressure is increased to 1330 torr, while the temperature is kept constant?
35. If a gas occupies 125 mL at 23.0°C, what volume will it occupy at 210°C, if the pressure is kept constant?
36. If a gas occupies 15.0 mL at 20.0°C and 7.00×10^2 torr, what volume will it occupy at 50.0°C and 6.00×10^2 torr?
37. A gaseous mixture containing oxygen, nitrogen, carbon monoxide, and carbon dioxide has a pressure of 2.00 atm. What is the partial pressure of the oxygen if $P_{nitrogen}$ = 532 mm Hg, $P_{carbon\ monoxide}$ = 684 mm Hg and $P_{carbon\ dioxide}$ = 76 mm Hg?
38. A 2.0 L sample of oxygen exerts a pressure of 480 torr, and a 2.0 L sample of nitrogen exerts a pressure of 0.35 atm. What would be the total pressure if these two

samples were combined in the same 2.0 L container?

39. A student collected a sample of hydrogen gas in a 1-liter container by water displacement. In her notebook, she recorded the barometric pressure as 756.8 mm Hg and the water temperature as 25°C.

    (a) What is the partial pressure of the hydrogen gas that she collected?

    (b) If she recorded the volume of the collected gases as 350 mL, what volume would the dry hydrogen occupy at STP?

Answers to the Check Your Understanding Questions in Chapter 3

(Answers have been rounded to the correct number of significant figures)

1. F **2.** T **3.** T **4.** F **5.** F **6.** F **7.** T **8.** T **9.** F **10.** F **11.** F **12.** T

13. (a) endothermic (b) endothermic (c) exothermic (d) exothermic (e) endothermic (f) endothermic

14. $1000 \text{ mL} \times \dfrac{1 \text{ g}}{1 \text{ mL}} \times \dfrac{539 \text{ cal}}{1 \text{ g}} = 539{,}000 \text{ cal} \times \dfrac{1 \text{ kcal}}{1000 \text{ cal}} = 539 \text{ kcal}$

15. 0.430 atm **16.** 157 torr **17.** 646 torr **18.** 1.450 atm **19.** 2100 mm Hg **20.** 7500 Pa

21. $V_2 = \dfrac{200 \text{ L} \times 755 \text{ mm Hg}}{370 \text{ mm Hg}} = 408 \text{ L}$

22. $P_2 = \dfrac{600 \text{ mL} \times 1.0 \text{ atm}}{80 \text{ mL}} = 7.5 \text{ atm}$

23. $P_2 = \dfrac{280 \text{ mL} \times 450 \text{ torr}}{1500 \text{ mL}} = 84 \text{ torr} \times \dfrac{1 \text{ atm}}{760 \text{ torr}} = 0.11 \text{ atm}$

24. $T_2 = \dfrac{25 \text{ mL} \times 310 \text{ K}}{250 \text{ mL}} = 31 \text{ K or } -242°\text{C}$

25. $V_2 = \dfrac{2.0 \text{ liters} \times 363 \text{ K}}{293 \text{ K}} = 2.5 \text{ liters}$

26. c **27.** b **28.** a **29.** a **30.** a **31.** a **32.** b **33.** c

34. $V_2 = \dfrac{115 \text{ mL} \times 763 \text{ mm Hg}}{1330 \text{ mm Hg}} = 66.0 \text{ mL}$

35. $V_2 = \dfrac{125 \text{ mL} \times 483 \text{ K}}{296 \text{ K}} = 204 \text{ mL}$

36. $V_2 = \dfrac{15.0 \text{ mL} \times 323 \text{ K} \times (7.00 \times 10^2) \text{ torr}}{293 \text{ K} \times (6.00 \times 10^2) \text{ torr}} = 19.3 \text{ mL}$

37. 228 mm Hg

38. $P_T = 746$ torr or 0.98 atm

39. (a) $P_{hydrogen} = 733.0$ mm Hg

(b) $\dfrac{733 \text{ mm Hg} \times 350 \text{ mL}}{298 \text{ K}} = \dfrac{760 \text{ mm Hg} \times V_2}{273 \text{ K}} = 309 \text{ mL}$

Chapter 4　　　　　ATOMIC STRUCTURE

In this chapter, we shall study the structure of the atom—the smallest unit of an element that has the properties of that element. There may be some vocabulary that is unfamiliar to you, so be sure that you understand the definitions of all the *Important Terms*.

4.1　　The Parts of the Atom

The atom is not indivisible, but is made up of many particles, the most important of which are protons, neutrons, and electrons. All of the positive protons and the neutral neutrons are found in the small dense nucleus at the center of the atom. The negative electrons are found in a relatively large region surrounding the nucleus.

Important Terms

　　　　proton　　　　neutron　　　　electron　　　　nucleus

4.2　　Atomic Number and Mass Number

The atomic number of an atom is equal to the number of protons in the nucleus of the atom. It is the number of protons that determines which element the atom represents. Each element has a specific and different atomic number. The number of electrons in a neutral atom is also equal to the atomic number. For example, the atomic number of sulfur is 16. Therefore, a neutral atom of sulfur will contain 16 protons and 16 electrons.

The mass number of an atom equals the number of protons plus the number of neutrons in its nucleus. Protons and neutrons have extremely small masses. Even so, they make up most of the mass of the atom because the mass of the electron is only 0.0005 times that of the proton or neutron. There are several shorthand ways to indicate the mass number and atomic number of an atom.

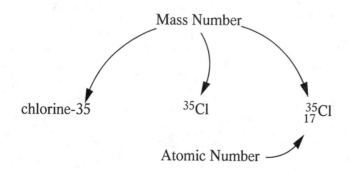

Important terms

atomic number   mass number

Example _____

State the number of protons, neutrons, and electrons in a neutral atom of radium-226.

S   Determine the number of neutrons, protons and electrons in a neutral atom of radium-226.

T   Use the table on the inside front cover of the text to find the atomic number of radium , which is also the number of protons. Then use the equation for the mass number to calculate the number of neutrons. The number of electrons in a neutral atom is the same as the number of protons.

E   The atomic number of radium is 88. Therefore, the number of protons and electrons is 88.

$$M = p + n$$
$$226 = 88 + n$$
$$138 = n$$

P   A neutral atom of radium-226 contains 88 protons, 88 electrons and 138 neutrons.

Check Your Understanding _____

For questions 1 to 6, fill in the blanks with the appropriate word or words.
1.   The nucleus of an atom contains _____ and _____.
2.   The identity of an atom is determined by the number of _____ in its nucleus.
3.   To figure out the number of electrons in a neutral atom, you need to know the

_____.
4.   The _____ of an atom equals the sum of the neutrons and protons in the atom's nucleus.
5.   The mass of a proton is greater than the mass of a(n) _____ and is the same as the mass of a(n) _____.
6.   The electron has a charge of _____, the proton a charge of _____, and the neutron a charge of _____.

7.   Give the number of protons, neutrons, and electrons in a neutral atom of each of the following:

(a) hydrogen-2   (b) ^{28}Si

(c) carbon-13

(d) oxygen-18

(e) $^{79}_{35}Br$

4.3   Isotopes

Often not all of the atoms of one element are alike. They must have the same number of protons in order to be atoms of the same element, but they can differ in the number of neutrons in their nuclei. Atoms of the same element that have different numbers of neutrons in their nuclei are called isotopes.

Important Term

isotope

4.4   Atomic Weight

Since atoms are so small, an arbitrary scale of relative weights has been established to measure them. In this scale, an atom of carbon-12 has been assigned a mass of exactly 12 atomic mass units (amu). Any sample of an element we might obtain from the world around us will contain a mixture of the isotopes of that element. As a result, the observed weight of an element is based on the relative abundance of the isotopes of that element. The atomic weight of an element is calculated by taking a weighted average of the masses of the isotopes of the element.

Important Terms

atomic weight   weighted average   atomic mass unit (amu)

Example _____

Gallium has two isotopes: gallium-68 and gallium-71. The relative abundance of gallium-68 is 76.76% and that of gallium-71 is 23.24%. Determine the atomic weight of gallium.

S   The question asks you to calculate the average atomic weight of gallium.

T   To calculate the atomic weight of gallium, multiply the mass of each isotope by its percent abundance and then add the products.

E

$$68 \times 76.76\% = 68 \times 0.7676 = 52.20$$
$$71 \times 23.24\% = 71 \times 0.2324 = \underline{16.50}$$
$$69.70 \text{ or } 70 \text{ amu}$$

P   The atomic weight of gallium to two significant figures is 70 amu.

Check Your Understanding ─────────────────────────────────

For questions 8 to 11, identify the statements as true (T) or false (F):
8.   Isotopes of an element differ in the number of protons in their nuclei.
9.   The actual mass of a carbon-12 atom is 12 grams.
10.   The atomic weight of an element is a weighted average of the masses of its isotopes.
11.   The atomic weight of an element with only one isotope equals the mass of the isotope.
12.   Use the data in Table 4.2 in the text to determine the atomic weight of carbon to two decimal places.

4.5   The Quantum Mechanical Model of the Atom

Since atoms are extremely small, scientists have used experimental data to develop elaborate theoretical models of the structure of the atom. Niels Bohr's model, based on the experimental data of Ernest Rutherford, pictured the atom as a small dense nucleus surrounded by electrons. The electrons cannot assume just any position around the nucleus, but rather occupy positions corresponding to specific energy levels. The energy of these levels increases as the electrons move farther from the nucleus. To move from one energy level to another, the electron must absorb or give off an amount of energy exactly equal to the difference in energy between the two energy levels. Each energy level is designated by a number, and can hold only a certain maximum number of electrons.

Because Bohr's model of the atom failed to explain some experimental data, a new model based on complicated mathematical calculations was developed. This model, called the quantum mechanical model, states that the position of an electron can never be determined exactly. Instead of specific orbits, the position of each electron is now described in terms of probability regions called orbitals. An orbital can contain zero, one, or two electrons.

Each energy level contains sublevels consisting of one or more atomic orbitals. The first energy level contains one orbital, the 1s orbital, whose probability distribution or shape is that of a sphere. The second energy level contains a 2s orbital and three 2p orbitals. The third energy level contains one 3s orbital, three 3p orbitals, and five 3d orbitals. The fourth energy level contains one 4s orbital, three 4p orbitals, five 4d orbitals, and seven 4f orbitals.

The rules for filling these orbitals are:

1.   An orbital can hold no more than two electrons, which must be spinning in opposite directions.

2.   An electron will not pair up with another electron if there is another orbital of the same energy available.

3.   The order of filling the orbitals is shown in Figure 4.4 in the text. Note that beginning with energy level 3 the energy levels overlap, so electrons begin filling a higher numbered level before the lower one is completely filled.

Important Terms

   Bohr model   energy level   quantum mechanical model

4.6   Electron Configurations

The arrangement of electrons or electron configuration of an atom can be shown in several ways. One is to list the orbitals in order of increasing energy and to use a superscript to show the number of electrons in each orbital. A second method called the orbital diagram uses a circle to indicate each orbital and an arrow for each electron.

Important Terms

   electron configuration   orbital diagram

Example   _____

1.   Write the orbital diagram for aluminum.

S   An orbital diagram like those in section 4.6 in the text should be drawn for aluminum.

T•E   A neutral atom of aluminum has 13 electrons. The 1s orbital is filled with 2 electrons. The 2s orbital is filled with 2 electrons. Each of the 2p orbitals holds 2 electrons, making a total of 6 electrons in the 2p orbitals. That makes 8 electrons in the second energy level, so it is now filled. We have placed 10 electrons in orbitals and have 3 left to assign to the third energy level. The 3s will be filled with 2 electrons and the last electron will be in a 3p orbital.

P $1s$ $2s$ $2p_x$ $2p_y$ $2p_z$ $3s$ $3p_x$ $3p_y$ $3p_z$

   (⥮) (⥮) (⥮) (⥮) (⥮) (⥮) (↿) ( ) ( )

2.   Write the electron configuration of aluminum.

S   The electron configuration for aluminum should be written using the same format as in section 4.6 of the text.

T•E   The solution to the previous question is used to write the electronic configuration.

P   Aluminum $1s^2 2s^2 2p^6 3s^2 3p^1$

Check Your Understanding ————————————————————————————

For questions 13 to 18, fill in the blanks with the correct word or words.

13.   The importance of the theory of Niels Bohr is that the electrons in the atom are restricted to definite energy positions called _____ or _____.
14.   Energy levels are identified by a _____.
15.   An electron in the fourth energy level has _____ (more, less) energy than an electron in the third energy level.
16.   For an electron to move from the second to the third energy level, it must _____ (gain, lose) energy.
17.   What is the maximum number of electrons in the first energy level? _____ The third energy level? _____ The fifth energy level? _____
18.   The arrangement of electrons in an atom is called the _____.

For questions 19 to 23, give the symbol and atomic number, and write the electron configurations for the elements.

19.   argon   22.   zinc
20.   vanadium   23.   tin
21.   cesium

4.7   Formation of Ions

If all the electrons in an atom are in the lowest possible energy levels, the atom is said to be in the ground state. Electrons in an atom can absorb energy from an exterior source and jump to a higher energy level. When this happens, the atom is said to be in an excited state. Atoms in the excited state are unstable and return to the ground state by giving off radiant energy. A positive ion is formed when one or more electrons absorb enough energy to jump completely away from the atom. Negative ions are formed when additional electrons are added to a neutral atom. Positive atoms are called cations and negative atoms are called anions.

Important Terms

          ground state   excited atom   ion
          cation   anion

55

For questions 24 to 26, fill in the blanks with the correct word or words.

24.   An atom can return to the ground state by _____ (gaining, losing) energy.

25.   When an atom absorbs enough energy for an electron to jump completely away from the atom, a _____ _____ is formed.

26.   A negative ion, called an _____, is formed when an atom _____ electrons; a positive ion, called a _____, is formed when an atom _____ electrons.

4.8   The Periodic Table

Nineteenth century scientists observed that regular, repeating characteristics existed among the known chemical elements, if these elements were arranged according to increasing atomic weight. Dmitri Mendeleev was the first to publish a table of elements showing the repeating nature, or periodicity, of the chemical properties of the elements. He predicted that the gaps in his table would be filled by elements not yet discovered. Mendeleev's table was refined by arranging the elements according to atomic number rather than atomic weight. The periodic table is a valuable tool to the scientist. It is important that you become familiar with its organization and be able to make use of the information it provides.

Important Terms

    periodic table   periodicity

4.9   Periods and Groups

The horizontal rows of the periodic table are called periods. The number of the period corresponds to the outermost energy level that contains electrons for the atoms of that period. Each vertical column is called a group or chemical family. The chemical behavior of an atom is determined by the number of electrons in the outermost energy level (the valence electrons.) Every member of a chemical family has the same number of valence electrons.

The eight chemical families called the representative elements are designated by the Roman numerals IA to VIIIA. These Roman numerals identify the number of valence electrons in the atoms of each member of that chemical family. Members of the same chemical family will exhibit very similar chemical behavior.

The three rows of ten elements in the center of the table (the B group elements) are called the transition elements or transition metals. The two rows of fourteen elements at the bottom of the table are called the inner transition elements.

Some groups or families of elements on the periodic table are known by common names:

group IA—alkali metals, group IIA—alkaline earth metals, and group VIIA—the halogens. The elements in group VIIIA, the noble gases, are unusual in their lack of chemical reactivity. This stability results from the number of valence electrons in these atoms (eight for each element except helium).

Important Terms

period	representative element	transition metal
group	inner transition element	alkali metal
halogen	alkaline earth metal	chemical family
noble gas	valence electron	

4.10   Metals, Nonmetals, Metalloids

The elements on the periodic table can be divided into two large classes: metals and nonmetals. The metals make up by far the majority of elements. A jagged line on the periodic table divides the metals from the nonmetals, but there is actually no sharp distinction between the properties of the elements immediately on either side of this line. These elements, called metalloids or semimetals, have properties of both metals and nonmetals.

Important Terms

Metal	Nonmetal	Metalloid	Semiconductor

Check Your Understanding ———————————————————————————

For questions 27 to 38, fill in the blanks with the correct word or words.
27.   The repeating nature of the chemical properties of the elements is called _____.
28.   The modern periodic table arranges the elements according to their _____.
29.   A vertical column on the periodic table is called a _____ or _____.
30.   A horizontal row on the periodic table is called a _____.
31.   The number of a row indicates _____.
32.   In the representative elements, the Roman numeral above a vertical column indicates

        _____.
33.   The electrons in the outermost energy level of an atom are called _____ electrons.
34.   Iodine (atomic number 53) will have _____ electrons in the outermost energy level, which is the _____ energy level.
35.   The representative elements are those in groups _____.
36.   Elements in group _____, called the noble gases, are chemically _____. This results from the number of their _____ electrons.
37.   Elements in the same chemical family will have similar _____.

38. An element that can be easily bent is most likely a _____, while one that is brittle is probably a _____.

Match items 39 to 48 with the correct area(s) on the following periodic table.

39. transition metals
40. nonmetals
41. hydrogen
42. alkali metals
43. inner transition elements
44. all metals
45. halogens
46. noble gases
47. metalloids
48. alkaline earth metals

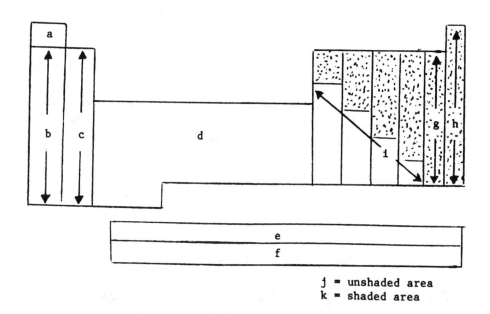

j = unshaded area
k = shaded area

4.11   The Periodic Law

The Periodic Law states that the physical and chemical properties of the elements will show a repeating nature or periodicity when the elements are arranged according to their atomic numbers. Three properties—atomic size, ionization energy, and electron affinity—all illustrate this periodicity. Understanding how these factors change on the periodic table will help you understand the chemical behavior of metals and nonmetals.

In general, the size of an atom decreases across a period and increases down a group. Metals tend to have large atomic radii, and nonmetals small radii. The ionization energy of an element increases as atomic number increases across a period, and decreases as atomic number increases down a family. Metals tend to have low ionization energies, while nonmetals have high ionization energies. The noble gases have unusually high ionization energies because of the very stable arrangement of their valence electrons. The electron affinities of the metals are low and the electron affinities of the nonmetals are high.

Important Terms

Periodic Law ionization energy electron affinity
atomic radius

Check Your Understanding ——————————————————————————————

To answer questions 49 to 51, use only the periodic table on the front inside cover of your text.
49.	For each of the following, choose the element with the largest radius.
	(a)	Se, K, Cr	(b)	Bi, N, As	(c)	Ge, F, Cs
50.	For each of the following, choose the element with the higher ionization energy.
	(a)	Li or Be	(c)	Br or Kr
	(b)	C or Sn	(d)	Mg or S
51.	Which will have the higher electron affinity, sodium or sulfur?

## 4.12	Elements Necessary for Life
Of the 90 naturally occurring elements, 25 are known to be essential to life. Of these 25, carbon, hydrogen, oxygen, and nitrogen are the most plentiful, making up 99.3% of all the atoms in your body. Although the remaining elements make up only 0.7% of the atoms in your body, they are critical to life. The lack of any one of them may cause disease or death.

We divide these 21 remaining elements into two groups: the macrominerals and the trace elements. The macrominerals include potassium, magnesium, sodium, calcium, phosphorus, sulfur, and chlorine. The trace elements, shown in Table 4.6, are found in very small amounts in the body. There is a specific range of concentration for each trace element that allows the body to function normally. Below that range a deficiency disease can develop. Concentrations above the range can be toxic.

Important Terms

macromineral trace element deficiency disease

Check Your Understanding ——————————————————————————————

For questions 52 to 54, fill in the blanks with the correct word or words.
52.	The four elements that are most abundant in living organisms are _____, _____, _____, and _____.

53. The next most abundant group of elements in the body contains _____ elements. They are called the _____.

54. The _____ elements exist in living organisms in minute amounts. These elements are made up of _____ metals and _____ nonmetals.

Review Your Understanding ───────────────────────────────────

For questions 55 to 64 select the best answer.

55. The atomic weights of the elements are compared to the atomic weight of:
    (a) hydrogen-1   (c) oxygen-16
    (b) nitrogen-14   (d) carbon-12

56. An atom containing 51 neutrons and 40 protons is an isotope of:
    (a) antimony (Sb)   (c) sodium(Na)
    (b) zirconium(Zr)   (d) protactinium(Pa)

57. If the atomic number of an element is 24 and its atomic mass is 52, the number of neutrons in its nucleus is:
    (a) 24   (c) 52
    (b) 28   (d) 76

58. What is the atomic mass of a krypton (Kr) atom that has 48 neutrons in its nucleus?
    (a) 12   (c) 36
    (b) 48   (d) 84

59. Which of the following is the ground state electronic configuration potassium (K)?
    (a) $1s^22s^22p^63s^23p^64s^1$
    (b) $1s^22s^22p^63s^23p^63d^1$
    (c) $1s^22s^22p^63s^23p^64p^1$
    (d) $1s^22s^22p^63s^23p^7$

60. Which of the following pairs are in the same group on the periodic table?
    (a) aluminum (Al) and sodium (Na)
    (b) sulfur (S) and polonium (Po)
    (c) potassium (K) and calcium (Ca)
    (d) thorium (Th) and uranium (U)

61. Which of the following pairs are in the same period on the periodic table?
    (a) nickel (Ni) and silver (Ag)
    (b) sodium (Na) and cesium (Cs)
    (c) silicon (Si) and tin (Sn)
    (d) iodine (I) and strontium (Sr)

62. The elements with atomic numbers 11 through 18 are classified as:
    (a) transition metals   (c) inner transition metals
    (b) representative elements   (d) noble metals

63. The element in group VIIA and the fourth period is:
    (a)   lead   (c)   Bromine
    (b)   chlorine   (d)   Iodine
64. The maximum number of electrons allowed in the third energy level is:
    (a)   2   (c)   8
    (b)   18   (d)   14

Answers to Check Your Understanding Questions in Chapter 4

1. protons, neutrons **2.** protons **3.** atomic number **4.** mass number **5.** electron, neutron
6. 1-, 1+, 0 **7.** (a) 1, 1, 1 (b) 14, 14, 14 (c) 6, 7, 6 (d) 8, 10, 8 (e) 35, 44, 35 **8.** F **9.** F **10.** T
11. T **12.** 12.01 amu **13.** energy levels, energy shells **14.** number **15.** more **16.** gain
17. 2, 18, 50 **18.** electron configuration
19. Ar, 18, $1s^2 2s^2 2p^6 3s^2 3p^6$
20. V, 23, $1s^2 2s^2 2p^6 3s^2 3p^6 4s^2 3d^3$
21. Cs, 55, $1s^2 2s^2 2p^6 3s^2 3p^6 4s^2 3d^{10} 4p^6 5s^2 4d^{10} 5p^6 6s^1$
22. Zn, 30, $1s^2 2s^2 2p^6 3s^2 3p^6 4s^2 3d^{10}$
23. Sn, 50, $1s^2 2s^2 2p^6 3s^2 3p^6 4s^2 3d^{10} 4p^6 5s^2 4d^{10} 5p^2$
24. losing **25.** positive ion **26.** anion, gains, cation, loses **27.** periodicity **28.** atomic numbers
29. group, family **30.** period
31. the number of the outermost energy level containing electrons
32. the number of electrons in the outermost energy level **33.** valence **34.** 7, 5th
35. groups IA-VIIIA **36.** VIIIA, unreactive, valence **37.** chemical properties
38. metal, nonmetal **39.** d **40.** k **41.** a **42.** b **43.** e, f **44.** j **45.** g **46.** h **47.** i **48.** c
49. (a) K (b) Bi (c) Cs **50.** (a) Be (b) C (c) Kr (d) S **51.** sulfur
52. carbon, hydrogen, oxygen, nitrogen **53.** 7, macrominerals **54.** trace, 8, 6
55. d, **56.** b, **57.** b, **58.** d, **59.** a, **60.** b, **61.** d, **62.** b, **63.** c, **64.** b

Chapter 5 COMBINATIONS OF ATOMS

In Chapter 4 we stated that the outer-shell electrons determine how atoms will interact with one another. In this chapter we examine these interactions between atoms. Atoms of elements in groups IA-VIIA will accept, share, or donate outer-shell electrons in order to attain eight valence electrons and become more stable. In doing so, these atoms form chemical bonds, the subject of this chapter.

5.1 The Octet Rule

The representative elements, those in Groups IA through VIIA, share or transfer electrons in order to form chemical combinations. These elements share or transfer exactly the number of electrons needed to give them the same electron arrangement as a noble gas. That is, they attain eight electrons in their outer shells.

Important Term

     octet rule

5.2 The Ionic Bond

An ionic bond results from the force of attraction between oppositely charged ions formed when electrons are transferred from one atom to another. Ionic bonds form between atoms that strongly attract additional electrons (the nonmetals) and atoms that have weak attraction for their valence electrons (the metals). These atoms gain or lose electrons to form a stable octet of electrons in their outer energy level.

For example, the metal lithium will react with bromine to form the ionic compound lithium bromide. Lithium needs to lose one electron to reach eight electrons in its outer shell, and bromine needs to gain one electron.

An ionic compound does not exist as separate molecules, but as groups of ions attracted to one another in an orderly arrangement called a crystal lattice. Ionic compounds are electrically neutral; the number of positive and negative charges in the lattice must be equal. For example, a crystal of lithium bromide will contain an equal number of positive lithium ions (Li^+) and negative bromide ions (Br^-).

Important Terms

    ionic bond   ionic compound   crystal lattice

5.3   Lewis Electron Dot Diagrams

The electron dot diagram is an easy way to represent the number of valence electrons in atoms of the representative elements. In these diagrams each dot represents a valence electron.

Important Term

electron dot diagram

Example _____

1. Write the electron dot diagram for sulfur.

S   The electron dot diagram for sulfur, S = (?)

T   From the inside front cover of the text we see that sulfur is in group VIA.

E   Sulfur therefore has six valence electrons.

P   The electron dot diagram for sulfur is $: \overset{\bullet}{\underset{\bullet}{S}} :$

2.   Write the electron dot diagram for the reaction between strontium and oxygen.

S   To answer this problem you must draw the electron dot diagrams for strontium and oxygen and for the ionic compound formed by their reaction.

T•E   From the inside front cover of the text, we see that the symbol for strontium is Sr. It is in group IIA and has two valence electrons. The symbol for oxygen is O and it is in group VIA. It has six valence electrons. Strontium must lose two electrons and oxygen must gain two electrons in order for both atoms to achieve an octet in their outer shell.

P

$$Sr \cdot + \cdot \overset{\bullet}{\underset{\bullet \bullet}{O}} : \longrightarrow Sr^{2+} + : \overset{\bullet \bullet}{\underset{\bullet \bullet}{O}} :^{2-}$$

Check Your Understanding _____

1.   Write the electron dot diagrams for the following elements:

(a)   sodium   (d)   selenium
(b)   calcium   (e)   bromine
(c)   gallium   (f)   argon

63

(g)   carbon   (h)   oxygen

2.   Use Lewis electron dot diagrams to show the reaction between the following
elements:
          (a)   calcium and oxygen   (b)   calcium and bromine

For questions 3 to 6, fill in the blanks with the correct word or words.
3.   _____ bonds result when metals and nonmetals interact.
4.   An _____ is formed by a group of positive and negative ions.
5.   When forming an ionic bond, metal atoms will _____ electrons, and nonmetal
atoms will _____ electrons.
6.   Metals form _____ ions and nonmetals form _____ ions.

5.4   The Covalent Bond

Ionic bonds form between elements that have large differences in their ability to attract
electrons. Such a transfer of electrons cannot take place between two elements that have
similar attraction for electrons (for example, two nonmetals). To become more stable, atoms
of these elements share electrons, forming covalent bonds. Covalently bonded atoms form
distinct units called molecules. The atoms of some nonmetallic elements do not exist
separately, but are found in covalently bonded pairs called diatomic molecules. In a bond
diagram, the covalent bond between the atoms is represented by a dash.

Important Terms

      covalent bond   covalent compound
      diatomic molecule   bond diagram

5.5   Multiple Bonds

Sometimes, atoms must share more than two electrons to achieve a stable octet in their
valence shell. A single covalent bond consists of two electrons (one pair) shared between
two nuclei. A double covalent bond contains four electrons (two pairs) shared between two
nuclei, and a triple covalent bond consists of six electrons (three pairs) shared between two
nuclei.

Important Terms

      single covalent bond   double covalent bond
      triple covalent bond

Example _____

Write the electron dot diagram and bond diagram for the compound formed between hydrogen and bromine.

S   You are asked to draw the electron dot and bond diagrams of the compound formed between hydrogen and bromine.

T•E   Both elements need to share one electron to become stable.

P   The electron dot diagram and bond diagram are:

          H : Br   H-Br

      Electron dot diagram   Bond diagram

Check Your Understanding _____

For questions 7 to 12, fill in the blanks with the correct word or words.

7.   In a covalent bond, electrons are _____ between two atoms.

8.   _____ bonds form between a metal and a nonmetal, while _____ bonds form between two nonmetals.

9.   Covalent compounds are composed of distinct units called _____.

10.   Six electrons shared between two atoms is called a _____ covalent bond, and is represented by _____ drawn between the symbols of the elements.

11.   A _____ covalent bond is two electrons shared between two atoms and is represented by _____ drawn between the symbols of the elements.

12.   Four electrons shared between two atoms is a _____ covalent bond and is represented by _____ drawn between the symbols of the elements.

13.   Write the formula, electron dot diagram, and bond diagram for the covalent compounds formed between the following elements:
          (a)   carbon and iodine   (c)   carbon and oxygen
          (b)   nitrogen and bromine   (d)   iodine and iodine

5.6   Electronegativity

Electrons in a covalent bond are not necessarily shared equally between the two nuclei. The tendency of an atom to attract shared electrons is called the electronegativity of that element. The more electronegative an element, the greater its attraction for shared electrons. On the periodic table, electronegativity increases across a period from left to right and decreases down a group.

Identical atoms will share electrons equally and will form nonpolar covalent bonds. Elements whose electronegativities differ will form polar covalent bonds, with the positive end of the bond nearer the less electronegative atom and the negative end of the bond nearer the more electronegative atom. If the electronegativity difference between the two atoms is great enough, the electrons will not be shared, but will be transferred—resulting in the formation of an ionic bond. There is no distinct separation between polar covalent bonds and ionic bonds. This is shown in Figure 5.7 in the text. Notice that there is a continuous range from nonpolar covalent compounds to ionic compounds. For our purposes it is convenient to make the following generalization: ionic bonds form between metals and nonmetals, and covalent bonds form between nonmetals.

Important Terms

electronegativity   polar covalent bond   nonpolar covalent bond

Check Your Understanding _____

For questions 14 to 17, indicate (a) which element is the more electronegative, and (b) the type of bond (ionic, nonpolar covalent, polar covalent) that will form between elements.
14.   chlorine and magnesium   16.   iodine and iodine
15.   phosphorus and chlorine   17.   bromine and potassium

5.7   Polar and Nonpolar Molecules
A molecule is nonpolar when the centers of positive and negative charge coincide. It is polar when these centers do not coincide. Whether a molecule is polar or nonpolar is determined both by the type of bonds present in the molecule (polar or nonpolar) and the arrangement of these bonds (the shape of the molecule). A molecule cannot be polar if all of its bonds are nonpolar. If the compound has polar bonds, it may be polar or nonpolar depending on its shape.

The nonmetallic elements in groups IVA to VIIA, when singly bonded to other elements, take on characteristic shapes. These shapes are illustrated in Figure 5.9 in the text.

Important terms

nonpolar molecule   polar molecule

Check Your Understanding _____

For questions 18 to 22, indicate (a) the type of bonding (polar covalent, nonpolar covalent, or ionic) found in the compound, and (b) if the bonding is covalent, indicate whether the molecule is polar or nonpolar.

18. CCl_4 19. NBr_3 20. LiBr   21. $CHBr_3$ 22. H_2

5.8   Hydrogen Bonding

A hydrogen bond is an attraction that forms between a hydrogen atom bonded to a highly electronegative atom on one molecule and a highly electronegative atom on another molecule (or on another region of the same molecule). Hydrogen bonds will only form between a hydrogen bonded to fluorine, nitrogen or oxygen and another fluorine, nitrogen or oxygen. A hydrogen bond is a weak attraction, equal to only about one-tenth that of an ordinary covalent bond. However, hydrogen bonds are critical in influencing the behavior of many biologically important molecules.

Important Terms

    hydrogen bond   intermolecular   intramolecular

Check Your Understanding ─────────────────────────────────

23.   In which three of the following might a hydrogen bond form?
        (a)   When a hydrogen atom is bonded to an iodine atom.
        (b)   When a hydrogen atom is bonded to an oxygen atom.
        (c)   When a hydrogen atom is bonded to a nitrogen atom.
        (d)   Between a hydrogen atom on one molecule and a hydrogen atom on another molecule.
        (e)   Between an oxygen atom on one molecule and a hydrogen atom on another molecule.

5.9   Polyatomic Ions

A polyatomic ion is a charged group of atoms held together as a stable unit by covalent bonds. This unit remains intact through most chemical reactions. Keep this in mind when writing the formulas of compounds that contain polyatomic ions. Polyatomic ions will form ionic bonds with metals. Table 5.6 in the text lists some common polyatomic ions. Your instructor will tell you which ones you should commit to memory.

Important Term

    polyatomic ion

5.10 Chemical Formulas

The chemical formulas of covalent compounds tell us what atoms are present and the number of each atom present in the compound. Formulas for ionic substances tell what ions are present and the ratio of these ions in the compound. In ionic compounds, the symbol of the positive ion is written first. If a chemical formula contains more than one polyatomic ion, the symbol for the polyatomic ion is placed in parentheses, followed by a subscript showing the number of polyatomic ions. For example, $Ca_3(PO_4)_2$.

5.11 Naming Chemical Compounds

Binary ionic compounds are named by placing the name of the positive ion first and the name of the negative ion second. The positive ion carries the name of the parent element. If the metal forms more than one positive ion, Roman numerals are placed after the name of the metal to show the charge on the ion. The negative ion uses the name of the parent element with an -ide suffix. Ionic compounds containing a metal and a polyatomic ion are named by placing the name of the metal first and the name of the polyatomic ion second.

Binary covalent compounds are named by writing the name of the element with the lower electronegativity first. The second element is named by adding the suffix -ide to the name of the parent element. Prefixes are used to show the number of atoms of each element present in the molecule.

Example　＿＿＿＿＿＿＿＿＿＿＿＿＿＿＿＿＿＿＿＿＿＿＿＿＿＿＿

1. Write the names of the following compounds:

T　　These problems require the *T* of our *STEP* method to arrive at the answer.

　　(a)　$ZnCl_2$　This is an ionic compound formed between the metal zinc and the nonmetal chlorine. The name of the metal ion is placed first with the nonmetal ion name ending with the suffix -ide: zinc chloride.

　　(b)　CuS　Note in Table 5.3 that copper can form two ions: Cu^{1+} and Cu^{2+}. To determine which copper ion this is, we use the fact that sulfur requires two electrons to become stable and there is only one copper ion. Therefore, copper must be donating two electrons. This compound must contain the copper (II) ion, Cu^{2+}. The name of this compound is copper(II) sulfide.

　　(c)　CS_2　This is a binary covalent compound. Sulfur is the more electronegative atom, so carbon is named first. There are two sulfur atoms, so the prefix "di" is used. The name is carbon disulfide.

2. Write the chemical formula for each of the following:

   (a)   **Barium chloride**: From Table 5.2, we see that barium forms a Ba^{2+} ion and chlorine forms a Cl^- ion. To maintain neutrality, there must be two chlorine ions for every barium ion. The formula is $BaCl_2$.

   (b)   **Aluminum oxide**: Again from Table 5.2, aluminum will form an Al^{3+} ion and oxygen an O^{2-} ion. To maintain neutrality, we need two aluminum ions (a total charge of 6+) and three oxide ions (a total charge of 6-).
   Using the crisscross method to determine the formula, we have

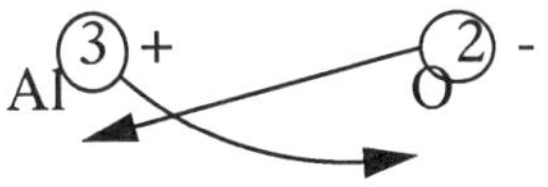

$$ Al^{\;\textcircled{3}+} \;\diagdown\diagup\; O^{\;\textcircled{2}-} \quad = \quad Al_2O_3 $$

   (c)   **Calcium nitrate**: This is an ionic compound formed between the calcium ion (Ca^{2+}) and the polyatomic nitrate ion (NO_3^-). Using the crisscross method, we have

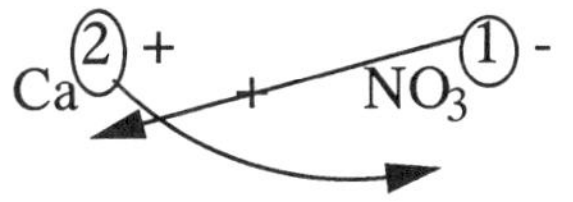

$$ Ca^{\;\textcircled{2}+} \;\diagdown\diagup\; NO_3^{\;\textcircled{1}-} \quad = \quad Ca(NO_3)_2 $$

Check Your Understanding ─────────────────────────────────────

For questions 24 to 43, write the chemical formula for each compound.

24.	silver sulfide	34.	sodium fluoride
25.	nitrogen dioxide	35.	carbon tetrachloride
26.	copper(I) acetate	36.	ammonium chloride
27.	aluminum sulfate	37.	iron(II) sulfite
28.	iron(III) oxide	38.	zinc bromide
29.	magnesium nitride	39.	aluminum oxide
30.	dinitrogen pentoxide	40.	hydrogen selenide
31.	potassium dichromate	41.	magnesium hydroxide
32.	nitrogen trichloride	42.	chlorine trifluoride
33.	barium sulfide	43.	copper(I) oxide

For questions 44 to 63, name each compound.

44. CuCl	49. NaI	54. HBr	59. SF_6
45. SO_3	50. N_2O_4	55. Al_2S_3	60. Li_2O
46. K_2S	51. $NaHSO_3$	56. CuO	61. $(NH_4)_2SO_4$
47. $FeBr_2$	52. $Mg_3(PO_4)_2$	57. H_2S	62. $LiClO_3$
48. SiO_2	53. KCN	58. $CaCl_2$	63. Na_2O_2

Review Your Understanding _____

64. The most common ion of strontium is:
    (a) Sr^+ (c) Sr^-
    (b) Sr^{2+} (d) Sr^{2-}

65. Sodium phosphate is:
    (a) $NaPO_4$ (c) Na_2PO_4
    (b) $Na(PO_4)_3$ (d) Na_3PO_4

66. Copper (I) bromide is:
    (a) CuBr   (c) Cu_2Br
    (b) $CuBr_2$ (d) Cu_2Br_2

67. Calcium hydroxide is:
    (a) Ca(OH)   (c) $Ca(OH)_2$
    (b) Ca_2OH (d) $CaOH_2$

68. Which of the following formulas is incorrect?
    (a) magnesium nitride, Mg_3N_2
    (b) potassium sulfate, K_2SO_4
    (c) ammonium phosphate, NH_4PO_4
    (d) sodium carbonate, Na_2CO_3

69. Which of the following is most likely to involve ionic bonding?
    (a) SO_2 (c) SCl_2
    (b) H_2 (d) Na_2S

70. Polar covalent bonding can be found in which of the following?
    (a) H_2 (c) K_2O
    (b) O_2 (d) NO_2

71. Magnesium and sulfur react to form magnesium sulfide. What is the formula for this substance?
    (a) MgS   (c) Mg_2S
    (b) MgS_2 (d) Mg_2S_2

72. Calcium reacts with bromine to form calcium bromide. What is the formula for this substance?
    (a) CaBr   (b) Ca_2Br

70

    (c) $CaBr_2$ (d) Ca_2Br_2

73.   The name of $Sr(NO_3)_2$ is:
    (a)   strontium nitride   (c)   strontium nitrate
    (b)   strontium dinitrate   (d)   strontium nitrite

Answers to Check Your Understanding Questions in Chapter

1. (a)   Na·   (b) · Ca ·   (c)· Ga ·   (d) : S̈e :   (e) : B̈r :

    (f) : A̤r ·   (h) : Ö ·

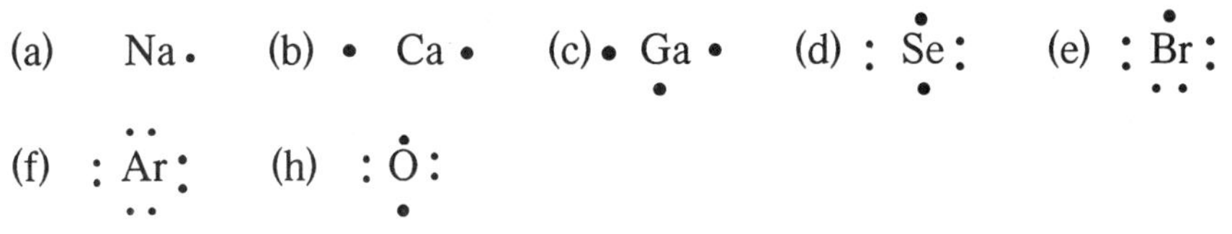

2. (a) · Ca · + ·Ö : ⟶ Ca^{2+} : Ö : $^{2-}$

    (b) : B̈r · + · Ca · + : B̈r · ⟶ :B̈r : $^-$ + Ca^{2+} + : B̈r: $^-$

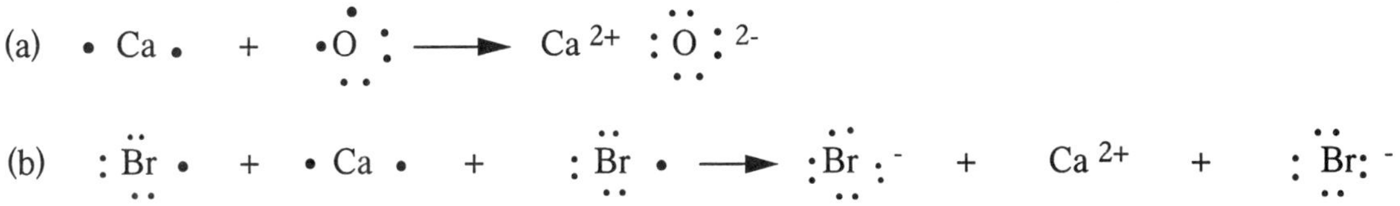

3. ionic 4. ionic compound (crystal lattice) 5. lose, gain 6. positive, negative 7. shared 8. ionic, covalent 9. molecules 10. triple, three dashes 11. single, dash 12. double, two dashes

13.
    (a) CI_4

$$: \ddot{I} :$$
$$: \ddot{I} : \overset{..}{C} : \ddot{I} :$$
$$: \ddot{I} :$$

$$\begin{array}{c} I \\ | \\ I-C-I \\ | \\ I \end{array}$$

    (b) NBr_3

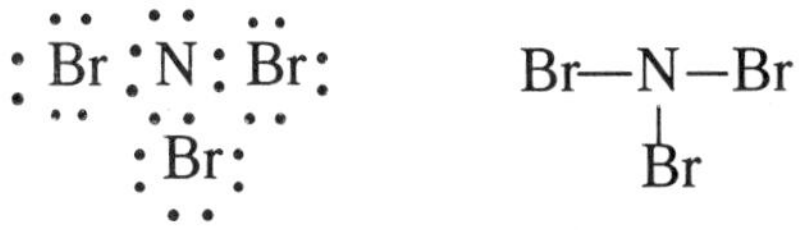

: B̈r : N : B̈r :
: B̈r :

$$\begin{array}{c} Br-N-Br \\ | \\ Br \end{array}$$

    (c) CO_2

·Ö:: C :: Ö·

O=C=O

    (d) I_2

: Ï: Ï:

I—I

14. (a) chlorine (b) ionic **15.** (a) chlorine (b) polar covalent **16.** (a) the same (b) nonpolar covalent **17.** (a) bromine (b) ionic **18.** (a) polar covalent (b) nonpolar **19.** (a) polar covalent (b) polar **20.** (a) ionic **21.** (a) polar covalent (b) polar **22.** (a) nonpolar covalent (b) nonpolar **23.** b,c,e **24.** Ag_2S **25.** NO_2 **26.** $CuC_2H_3O_2$ **27.** $Al_2(SO_4)_3$ **28.** Fe_2O_3 **29.** Mg_3N_2 **30.** N_2O_5 **31.** $K_2Cr_2O_7$ **32.** NCl_3 **33.** BaS **34.** NaF **35.** CCl_4 **36.** NH_4Cl **37.** $FeSO_3$ **38.** $ZnBr_2$ **39.** Al_2O_3 **40.** H_2Se **41.** $Mg(OH)_2$ **42.** ClF_3 **43.** Cu_2O
44. copper(I) chloride **45.** sulfur trioxide **46.** potassium sulfide **47.** iron(II) bromide
48. silicon dioxide **49.** sodium iodide **50.** dinitrogen tetroxide **51.** sodium hydrogen sulfite
52. magnesium phosphate **53.** potassium cyanide **54.** hydrogen bromide
55. aluminum sulfide **56.** copper(II) oxide **57.** hydrogen sulfide **58.** calcium chloride
59. sulfur hexafluoride **60.** lithium oxide **61.** ammonium sulfate **62.** lithium chlorate
63. sodium peroxide **64.** b **65.** d **66.** a **67.** c **68.** c **69.** d **70.** d **71.** a **72.** c **73.** c

Chapter 6   CHEMICAL EQUATIONS AND THE MOLE

In this chapter we discuss writing and balancing equations that represent chemical reactions. A second important concept introduced is the mole, the reacting unit used in chemistry. By the time you have finished this chapter, it is important that you feel comfortable with the definition of the mole and calculations using the mole.

6.1   Writing Chemical Equations

Chemical equations are a shorthand method of describing a chemical reaction. The substances that you begin with are called the reactants, and appear to the left of the arrow in the chemical equation. The substances formed in a chemical reaction are called the products, and appear to the right of the arrow in the chemical equation.

When chemical reactions occur, atoms are neither created nor destroyed—only rearranged. Therefore, the number of atoms of each element on the reactant side of the equation must equal the number of atoms of that element on the product side of the equation. An equation written this way is said to be balanced. In a chemical reaction, the total mass of the reactants will equal the total mass of the products (the Law of Conservation of Mass).

Important Terms

> reactant   product
> Law of Conservation of Mass   balanced equation

6.2   Balancing Chemical Equations

Study carefully the four steps described in your text for balancing chemical equations. You might want to refer to them when reading the following examples and then try each example on your own. Confidence in balancing chemical equations comes only with lots and lots of practice.

Example

1.   Magnesium reacts with oxygen to produce magnesium oxide, write a balanced chemical equation for this reaction.

S   Balance the equation. The reactants for this chemical reaction are elemental magnesium and oxygen. The product is magnesium oxide.

$$Mg \ + \ O_2 \longrightarrow \quad MgO$$

T•E Looking at the above equation, we see that the magnesium atoms balance, but not the oxygen atoms. In order to have the same number of oxygen atoms on both sides of the equation we must have two magnesium oxides. (Remember that once you have written the correct chemical formulas for the reactants and products you cannot change those formulas, you can only adjust the coefficients to balance the equation.)

$$Mg \ + \ O_2 \longrightarrow \ 2 \ MgO$$

Now the oxygen atoms balance, but not the magnesium atoms. Since there are two magnesium atoms on the product side, there must be two magnesium atoms on the reactant side.

$$2 \ Mg \ + \ O_2 \longrightarrow \ 2 \ MgO$$

P There are two magnesium atoms on each side and two oxygen atoms on each side, so the equation is balanced.

2. Sodium hydroxide will react with iron(III) chloride to produce iron(III) hydroxide and sodium chloride.

S Write a balanced equation for this reaction. The reactants in this case are sodium hydroxide and iron(III) chloride. The products are iron(III) hydroxide and sodium chloride.

T•E Sodium is in group IA and forms a 1+ ion. Iron(III) is the ion formed by iron which has a 3+ charge. Chloride is in group VIIA and forms a 1- ion. The hydroxide ion is listed in Table 5.6 in the text. It is critical that you spend time to be sure that each chemical formula is correct because the equation cannot be balanced properly if a formula is wrong.

$$NaOH \ + \ FeCl_3 \longrightarrow \quad Fe(OH)_3 \ + \quad NaCl$$

Begin by giving $Fe(OH)_3$ a coefficient of 1. The iron atoms balance. There are three hydroxides (OH) on the product side, so we need three hydroxide ions on the reactant side.

$$3 \ NaOH \ + \ FeCl_3 \longrightarrow \quad Fe(OH)_3 \ + \quad NaCl$$

Now balance the sodium atoms by putting a coefficient of three before the NaCl on the product side. Check the chlorine atoms. Do they balance?

$$3 \text{ NaOH} \ + \ \text{FeCl}_3 \ \longrightarrow \ \text{Fe(OH)}_3 \ + \ 3 \text{ NaCl}$$

P Check this equation yourself. Is it balanced?

3. Balance the following equation:

$$\text{Si} \ + \ \text{Cr}_2\text{O}_3 \longrightarrow \ \text{SiO}_2 \ + \ \text{Cr}$$

S The formulas for the reactants and products are already written.

T•E This is not a balanced equation, there are two chromiums on the left side of the equation, but only one on the right side. Placing the coefficient 2 in front of the Cr will balance chromium.

$$\text{Si} \ + \ \text{Cr}_2\text{O}_3 \longrightarrow \ \text{SiO}_2 \ + \ 2 \text{ Cr}$$

Now we must balance the oxygens. There are 3 oxygens on the left side and two on the right side. The smallest number that 2 and 3 both divide evenly into is 6. We need to have 6 oxygens on each side of the equation. If we place the coefficient 2 in front of the Cr_2O_3 and the coefficient 3 in front of the SiO_2, the oxygens will balance.

$$\text{Si} \ + \ 2 \text{ Cr}_2\text{O}_3 \ \longrightarrow \ 3 \text{ SiO}_2 \ + \ 2 \text{ Cr}$$

Now the chromiums are no longer balanced, there are 4 chromiums on the left side of the equation and only two on the right side. To rebalance the chromiums put the coefficient 4 in front of the Cr on the right side of the equation.

$$\text{Si} \ + \ 2 \text{ Cr}_2\text{O}_3 \ \longrightarrow \ 3 \text{ SiO}_2 \ + \ 4 \text{ Cr}$$

Next we must balance the silicons. There are 3 silicons on the right side of the equation and only 1 on the left side. Put the coefficient 3 in front of the Si.

$$3 \text{ Si} \ + \ 2 \text{ Cr}_2\text{O}_3 \ \longrightarrow \ 3 \text{ SiO}_2 \ + \ 4 \text{ Cr}$$

P Check to be sure that the equation balances.

Check Your Understanding ————————————————————————
For questions 1 to 5, write a sentence saying in words what is indicated by each chemical equation.

1. $\text{H}_2 \ + \ \text{S} \longrightarrow \ \text{H}_2\text{S}$
2. $\text{Ca} \ + \ 2 \text{ H}_2\text{O} \longrightarrow \ \text{Ca(OH)}_2 \ + \ \text{H}_2$
3. $\text{Cu} \ + \ 2 \text{ AgNO}_3 \longrightarrow \ 2 \text{ Ag} \ + \ \text{Cu(NO}_3)_2$
4. $2 \text{ H}_2\text{O} \ + \ \text{C} \longrightarrow \ \text{CO}_2 \ + \ 2 \text{ H}_2$
5. $\text{F}_2 \ + \ 2 \text{ NaCl} \longrightarrow \ 2 \text{ NaF} \ + \ \text{Cl}_2$

For questions 6 to 15, balance the equations.

6. $HCl + Fe \longrightarrow FeCl_2 + H_2$
7. $N_2 + O_2 \longrightarrow NO_2$
8. $H_3PO_4 + Ca(OH)_2 \longrightarrow Ca_3(PO_4)_2 + H_2O$
9. $Pb + Al_2O_3 \longrightarrow Al + PbO$
10. $Al + O_2 \longrightarrow Al_2O_3$
11. $CH_4 + O_2 \longrightarrow CO_2 + H_2O$
12. $Ni + As \longrightarrow Ni_3As_2$
13. $Al + N_2 \longrightarrow AlN$
14. $C_2H_2 + O_2 \longrightarrow CO_2 + H_2O$
15. $Mg + Fe_2O_3 \longrightarrow MgO + Fe$

For questions 16 to 20, write the chemical equation for the reaction and balance then balance the equation.

16. Potassium reacts with oxygen to form potassium oxide.
17. Tin reacts with bromine to form tin(II) bromide.
18. Magnesium reacts with nitrogen to form magnesium nitride.
19. Hydrogen sulfide reacts with oxygen to form sulfur dioxide and water.
20. Calcium hydroxide reacts with sodium carbonate to form sodium hydroxide and calcium carbonate.

6.3 Oxidation-Reduction Reactions

Chemical reactions that involve the transfer of electrons are called oxidation-reduction or redox reactions. Oxidation is the loss of electrons by an atom and reduction is the addition of these electrons to another atom. In redox reactions, the element that gains electrons is the oxidizing agent and the element that loses electrons is the reducing agent. In a chemical reaction the number of electrons lost must equal the number of electrons gained.

Important Terms

    oxidation   reduction   redox reaction
    oxidizing agent   reducing agent

6.4   The Mole

This section introduces a key concept—the reacting unit in chemistry—the mole. The mole allows scientists to control the exact amount of each element or compound used in the laboratory. Since atoms are so small, it is impossible to measure out one thousand or even one million atoms of an element from a reagent bottle. Therefore, an arbitrary unit was established that allows scientists to measure an exact number of atoms. This unit, the mole, contains 6.02×10^{23} units (the units themselves depend on the specific nature of the substance). One mole of potassium ions contains 6×10^{23} ions, one mole of water contains 6×10^{23} water molecules, one mole of magnesium contains 6×10^{23} atoms, and one mole of donuts contains 6×10^{23} donuts.

The weight of a mole depends upon the nature of the substance involved. For example, one mole or 6×10^{23} peas would weigh significantly less than one mole or 6×10^{23} watermelons. The weight of one mole of atoms of an element is defined as the atomic weight of that element in grams. Therefore, one mole of magnesium weighs 24 grams, and one mole of helium weighs four grams.

Note: In problems involving calculations with the mole, we will use atomic weights to three significant digits.

Important Terms

mole (mol)   Avogadro's number

Example _____

1.   Determine the weight in grams of 1.75 mol of selenium.

S   The problem asks: 1.75 mol = (?) g of Se.

T   From the inside front cover of the text, we learn that the atomic weight of selenium is 78.96 amu. From the definition of a mole, the weight of one mole is the atomic weight expressed in grams.

$$1 \text{ mol Se } = 79.0 \text{ g}$$

We can write two conversion factors from this equality.

$$\frac{1 \text{ mol Se}}{79.0 \text{ g}} \qquad \text{and} \qquad \frac{79.0 \text{ g}}{1 \text{ mol Se}}$$

E   We must now decide which conversion factor is the one to use. The first conversion factor converts grams to moles, while the second conversion factor converts moles to grams. Because the answer needs to be in grams, the second conversion factor is the one to use.

$$1.75 \text{ mol Se} \times \frac{79.0 \text{ g}}{1 \text{ mol Se}} = 138.25 \text{ g}$$

P   Now decide whether the answer from the calculations is correct. It is often useful to make a rough estimate of the expected answer. The weight of one mole of Se is about 80 grams and the problem asks for 1.75 moles. The answer should be less than 160 (roughly two moles) but more than 120 (roughly one and one-half moles). The answer fits within this range. Because the quantity 1.75 mol has three significant figures, the answer must have only three significant figures and is 138 g Se.

Check Your Understanding ———————————————————————

For questions 21 to 27, choose the correct answer.

21.   One mole is equal to the number of atoms in
     (a)   12.0000 grams of the carbon-12 isotope.
     (b) 6×10^{23} grams of the carbon-12 isotope
     (c) 6×10^{23} grams of any element
     (d)   12.000 grams of any element
22.   Avogadro's number refers to
     (a)   the weight of one mole of any substance
     (b)   the number of atoms in 1 gram of any element
     (c)   the number of atoms in a mole
     (d) 6×10^{23} grams of any substance
23.   The weight of one mole of any substance equals
     (a)   Avogadro's number
     (b)   the formula weight in grams
     (c)   the atomic weight in grams
     (d) 6×10^{23} grams
24.   The weight of one mole of potassium atoms is.
     (a)   78.2 g   (c)   31.0 g
     (b)   39.1 g   (d)   12.0 g

25. The weight of 2.80 moles of beryllium atoms is
    (a) 9.01 g   (c) 18.0 g
    (b) 27.3 g   (d) 25.2 g
26. The weight of 0.360 mole of argon atoms is
    (a) 14.4 g   (c) 40.0 g
    (b) 144 g   (d) 111 g
27. The weight of 3×10^{22} atoms of helium is
    (a) 2 g   (c) 0.2 g
    (b) 0.4 g   (d) 0.02 g

6.5   Formula Weight

The formula weight of a substance equals the sum of the atomic weights of all the atoms in the formula of that substance. The formula weight of chlorine (Cl_2) is $35.5 + 35.5 = 71.0$ amu. The formula weight of hydrochloric acid (HCl) is $1.0 + 35.5 = 36.5$ amu. The formula weight of calcium hydroxide [$Ca(OH)_2$] is calculated by adding the atomic weight of calcium to twice the sum of the atomic weight of hydrogen plus the atomic weight of oxygen: $40.1 + 2(1.0 + 16.0) = 40.1 + 34.0 = 74.1$ amu.

One mole of any substance has a mass equal to the formula weight of that substance in grams. One mole of chlorine weighs 71.0 grams, one mole of hydrochloric acid weighs 36.5 grams, and one mole of calcium hydroxide weighs 74.1 grams.

Important Term

   formula weight

Example   _____

Calculate the formula weight in grams of $Ca(NO_3)_2$.

S   The question asks: Formula weight of $Ca(NO_3)_2$ = (?) g.
T   The formula weight is the sum of the atomic weights of all of the atoms in the formula
E   From the periodic table we see that Ca = 40.1 amu, N = 14.0 amu, and O = 16.0 amu. Therefore:

Ca		= 40.1 amu
2 x N	2 x 14.0	= 28.0 amu
6 x O	6 x 16.0	= 96.0 amu
$Ca(NO_3)_2$		= 164.1 amu

P   The answer has the correct number of significant figures and should be expressed in grams as 164.1 g.

Check Your Understanding ————————————————————————
For questions 28 to 30, choose the correct answer.

28.   The formula weight of $BaCl_2$ is
    (a)   20.8 amu   (c)   81.8 amu
    (b)   172 amu   (d)   208 amu

29.   The formula weight of $Al_2(CO_3)_3$ is
    (a)   87.0 amu   (c)   234 amu
    (b)   114 amu   (d)   207 amu

30.   The weight of one mole of $Ca_3(PO_4)_2$ is
    (a)   310 g   (c)   135 g
    (b)   230   (d)   184 g

31.   What is the formula weight of each of the following?
    (a) MnO_2 (c) CO_2 (e) $KClO_3$ (g) H_2
    (b) $Al(NO_3)_3$ (d) $Be_3(PO_4)_2$ (f) H_2SO_4 (h) NaBr

6.6 Molar Volume of a Gas

One mole of an ideal gas occupies 22.4 L under conditions of standard temperature and pressure. This means that 32 g (1 mol) of O_2 or 28 g (1 mol) of N_2 or 20 g (1 mol) of Ne or 1 mole of any gas has a volume 22.4 L at STP. Because of this relationship, we can convert between the volume of a gas and its weight as in the following examples.

Important Terms

       Avogadro's law   molar volume

Example ————————————————————————————————

1.   What is the volume of .75 mol chlorine at STP?

S   The problem asks: Volume of chlorine = (?)

T   The volume of one mole of any gas at STP is 22.4 L. We can set up two conversion factors based on this relationship:

$$\frac{1 \text{ mol}}{22.4 \text{ L}} \quad \text{and} \quad \frac{22.4 \text{ L}}{1 \text{ mol}}$$

E   The second conversion factor is used to solve this problem.

$$0.75 \; \cancel{mol} \; x \; \frac{22.4 \, L}{1 \, \cancel{mol}} = 16.8 \, L$$

P   The answer must have two significant figures and is 17 L.

2.   What is the volume of a 28.5 g sample of argon at STP?

S   The problem asks: Volume of 28.5 g Argon = (?)

T   From the inside front cover of the text we find that the atomic weight of 1 mol of argon is 40.0 g. From this we can determine the number of moles of argon. We can then determine the volume of our sample using Avogadro's law. The conversion factors to use are:

$$\frac{1 \, mol \, Ar}{40.0 \, g} \quad and \quad \frac{22.4 \, L}{1 \, mol}$$

E

$$28.5 \; \cancel{g} \, Ar \; x \; \frac{1 \, \cancel{mol} \, Ar}{40.0 \, \cancel{g} \, Ar} \quad x \quad \frac{22.4 \, L}{1 \, \cancel{mol}} \; = \; 15.96 \, L$$

P   The measurement 28.5 g has three significant figures so the answer must have only thee significant figures and is 16.0 L

3.   If a sample of oxygen gas weighs 24.0 g at 25°C and 740 mm Hg, what volume will it occupy?

S   The problem asks for the volume of the sample of oxygen under the conditions of temperature and pressure given.

T   To solve this problem use Avogadro's law to determine the volume of 24.0 g of oxygen at STP and then use the combined gas law equation to convert the volume at STP to the volume at 25°C and 740 mm Hg.

$$24.0 \; \cancel{g} \; x \; \frac{1 \, \cancel{mol} \, O_2}{32.0 \, \cancel{g}} \; x \; \frac{22.4 \, L}{1 \, \cancel{mol}} = 16.8 \, L$$

$$16.8 \, L \; x \; \frac{298 \, K}{273 \, K} \; x \; \frac{760 \, \cancel{mm \, Hg}}{740 \, \cancel{mm \, Hg}} = 18.834 \, L \; (\, calculator \; answer)$$

P   The answer must have three significant figures and is 18.8 L. (Remember that 1 mol is a definition of a unit of measure and is not considered in determining how many significant figures there should be in the answer. The precision in the actual mass of a mole of a substance does need to be considered in deciding how many significant figures should appear in the answer.)

Check Your Understanding _____

32. A sample of carbon dioxide occupies a 0.500 L container at 170 torr and 25°C.
     (a)   What is the weight of this sample in milligrams?
     (b)   What would be the pressure of this sample if it were placed in a 65.0 mL tube at 15°C? (Hint: make sure that the units for the initial and final P, V, and T are the same.)

6.7   Using the Mole in Problem Solving

Doing calculations using the mole can be simplified if you use the conversion factor method discussed in Chapter 1. If you have not done so recently, reread Section 1.9. Use of the *STEP* method will help you solve these problems. The following steps are helpful to follow when solving problems involving the mole.

S   See the Question. Determine what the question is asking for.

T   Think it Through. Establish a relationship between the given quantity in the problem and the unknown quantity. (In some problems this may require more than one equation.) Write conversion factors from this equation (or equations).

E   Solve the problem, making certain to cancel units so that the answer is in the correct units.

P   Check the answer to make certain that it is in the desired units and that it has the correct number of significant figures.

Example _____

1.   How much does 0.0240 mole of calcium hydroxide weigh?

S $0.0240 \text{ mol Ca(OH)}_2 = ?$ grams.

T $1 \text{ mol Ca(OH)}_2 = 74.1 \text{ g}$

$$\frac{1 \text{ mol Ca(OH)}_2}{74.1 \text{ g}} \quad \text{and} \quad \frac{74.1 \text{ g}}{1 \text{ mol Ca(OH)}_2}$$

E $0.0240 \text{ mol Ca(OH)}_2 \times \dfrac{74.1 \text{ g}}{1 \text{ mol Ca(OH)}_2} = 1.78 \text{ g}$

P   Because 0.0240 mole has three significant figures, the answer 1.78 g has the correct number of significant figures.

2.   How many moles are there in 355 milligrams of chlorine gas?

S $355 \text{ mg} = ? \text{ mol } Cl_2$

T $1 \text{ mol } Cl_2 = 71.0 \text{ g}$ and $1 \text{ g} = 1000 \text{ mg}$ (Remember: chlorine exists as the diatomic molecule Cl_2)

$$\frac{1 \text{ mol } Cl_2}{71.0 \text{ g}} \quad \text{and} \quad \frac{71.0 \text{ g}}{1 \text{ mol } Cl_2} \qquad\qquad \frac{1 \text{ g}}{1000 \text{ mg}} \quad \text{and} \quad \frac{1000 \text{ mg}}{1 \text{ g}}$$

E

$$355 \text{ mg} \times \frac{1 \text{ g}}{1000 \text{ mg}} \times \frac{1 \text{ mol } Cl_2}{71.0 \text{ g}} = \frac{355 \text{ mol } Cl_2}{1000 \times 71.0}$$

$$= 0.00500 \text{ mol } Cl_2$$

P   The answer must have three significant figures and is $0.00500 \text{ mol } Cl_2$. This answer could also have been expressed as an exponential number $5.00 \times 10^{-3} \text{ mol } Cl_2$.

Check Your Understanding _____

For questions 33 and 34, you will want to refer to the formula weights that you determined in question 31.
33.   How many grams are there in each of the following?
    (a) $3.00 \text{ mol of } MnO_2$
    (b) 2.00×10^{21} atoms of $Al(NO_3)_3$
    (c) $0.250 \text{ mol of } CO_2$
    (d) $4.50 \text{ mol of } Be_3(PO_4)_2$
34.   How many moles are there in each of the following?
    (a) 2.40×10^{24} formula units of $KClO_3$
    (b)   0.482 grams of H_2SO_4
    (c)   1.00 kg of H_2
    (d)   4.60 mg of sodium bromide

6.8   Calculations Using Balanced Equations
A balanced equation tells us many things about a reaction. It tells us the number of atoms, the number of molecules of a covalent compound, the number of formula units of an ionic

compound, the number of moles, even the number of grams of each substance involved in the reaction. Look at the following balanced equation:

$$Zn + 2\,HCl \longrightarrow ZnCl_2 + H_2$$

This equation can be read in the following ways:

1. One atom of zinc will react with two molecules of hydrochloric acid to form one formula unit of zinc chloride and one molecule of hydrogen.

2. One mole of zinc will react with two moles of hydrochloric acid to form one mole of zinc chloride and one mole of hydrogen.

3. 65.4 grams of zinc will react with 73.0 grams of hydrochloric acid to form 136.4 grams of zinc chloride and 2.0 grams of hydrogen. (Do you see that statement 3 follows directly from statement 2?)

All this information allows us to perform calculations to determine unknown quantities in a reaction.

Example _____

1. Assume you have 13.0 grams of zinc. If you react all of it with hydrochloric acid, how many grams of zinc chloride will you produce?

S Using the balanced equation for the reaction,

$$Zn + 2\,HCl \longrightarrow ZnCl_2 + H_2$$

we need to determine how many grams of zinc chloride will be produced if 13.0 grams of zinc react.

T We need unit factors for the following conversions:

$$\overset{(1)}{\text{grams Zn} \longrightarrow} \overset{(2)}{\text{moles Zn} \longrightarrow} \overset{(3)}{\text{moles ZnCl}_2 \longrightarrow} \text{grams Zn}$$

(1) The first unit factor we need involves the conversion of grams of Zn to moles of Zn.

$$\frac{1 \text{ mol Zn}}{65.4 \text{ g Zn}} \quad \text{and} \quad \frac{65.4 \text{ g Zn}}{1 \text{ mol Zn}}$$

(2) For the second conversion we need to look at the ratio of coefficients in our balanced equation.

$$\frac{1 \text{ mol Zn}}{1 \text{ mole ZnCl}_2} \quad \text{and} \quad \frac{1 \text{ mol ZnCl}_2}{1 \text{ mol Zn}}$$

(3)   For the third conversion, we need a conversion factor showing the ratio of moles of zinc chloride to grams of zinc chloride.

$$\frac{1 \text{ mol ZnCl}_2}{136.4 \text{ g ZnCl}_2} \quad \text{and} \quad \frac{136.4 \text{ g ZnCl}_2}{1 \text{ mol ZnCl}_2}$$

E   To solve the problem in one step, we must choose the conversion factors from (1), (2), and (3) in such a way that all the units cancel except grams of $ZnCl_2$.

$$13.0 \text{ g Zn} \times \frac{1 \text{ mol Zn}}{65.4 \text{ g Zn}} \times \frac{1 \text{ mol ZnCl}_2}{1 \text{ mol Zn}} \times \frac{136.4 \text{ g ZnCl}_2}{1 \text{ mol ZnCl}_2}$$

$$= 27.11315 \text{ g ZnCl}_2 \text{ (calculator answer)}$$

P   The answer must be rounded to three significant figures and is 27.1 g

2.   What is the minimum number of grams of hydrochloric acid that must be present to react with all of the zinc in example 1?

S   The problem refers to the same balanced equation as in example 1. We must determine how many grams of hydrochloric acid will react with 13.0 g Zn.

T   From the balanced equation, we see that for every mole of zinc reacted, 2 moles of hydrochloric acid will react. If we know how many moles of zinc we have, we can calculate the required moles of HCl and grams of HCl.

E   The problem can be solved in one step by use of conversion factors.

$$13.0 \text{ g Zn} \times \frac{1 \text{ mol Zn}}{65.4 \text{ g Zn}} \times \frac{2 \text{ mol HCl}}{1 \text{ mol Zn}} \times \frac{73.0 \text{ g HCl}}{1 \text{ mol HCl}}$$

$$= 29.021407 \text{ g HCl (calculator answer)}$$

P   The answer must have three significant figures and is 29.0 g.

3.   How many grams of sulfur dioxide will be produced when 500 g of hydrogen sulfide is reacted with 700 g of oxygen? (The other product is water.)

S   We are asked to the determine how many grams of sulfur dioxide will be produced. First we must write a balanced equation for this reaction.

$$2 \text{ H}_2\text{S} + 3 \text{ O}_2 \longrightarrow 2 \text{ SO}_2 + 2 \text{ H}_2\text{O}$$

T   Often in the industrial production of chemicals, one of the reactants will be added in excess. The reactant that is completely reacted (and used up) before any other is called the limiting reactant. We will need to determine the limiting reactant to solve this problem.

E   We need to determine the number of moles of each reactant.

$$500 \ \cancel{g} \times \frac{1 \ \text{mol} \ H_2S}{34.1 \ \cancel{g}} \ = \ 15.6 \ \text{mol} \ \text{hydrogen sulfide}$$

$$700 \ \cancel{g} \times \frac{1 \ \text{mol} \ O_2}{32.0 \ \cancel{g}} \ = \ 21.8 \ \text{mol oxygen}$$

From the balanced equation, we see that three moles of hydrogen sulfide will react with two moles of oxygen, so for 15.6 mol of H_2S, we need

$$15.6 \ \cancel{\text{mol} \ H_2S} \times \frac{3 \ \text{mol} \ O_2}{2 \ \cancel{\text{mol} \ H_2S}} = 23.4 \ \text{mol oxygen}$$

Thus, we see that we don't have enough oxygen to react with all of the hydrogen sulfide. So oxygen will be the limiting reactant. Since the reaction will stop when all of the oxygen is used, we use the number of moles of the limiting reagent, oxygen, to calculate the number of grams of sulfur dioxide that will be produced.

$$21.8 \ \cancel{\text{mol} \ O_2} \times \frac{2 \ \cancel{\text{mol} \ SO_2}}{3 \ \cancel{\text{mol} \ O_2}} \ \times \ \frac{64.1 \ \text{g}}{1 \ \cancel{\text{mol} \ SO_2}} \ = \ 931.58667 \ \text{g oxygen}$$

P   The answer is 931 g O_2

Check Your Understanding _____

For questions 35 to 37, write the equations as sentences involving (a) atoms, molecules, and formula units; (b) moles; and (c) grams.

35. $N_2 + 3 H_2 \longrightarrow 2 NH_3$
36. $FeCl_3 + 3 NaOH \longrightarrow Fe(OH)_3 + 3 NaCl$
37. $F_2 + 2 NaCl \longrightarrow 2 NaF + Cl_2$

38.   How many grams of water are formed when 51.0 grams of ammonia are burned in oxygen? (The other product is nitrogen gas.)

39.   Chlorine is a very important industrial chemical. Many tons of chlorine are produced each year by the breakdown of sodium chloride. The other product of the reaction is sodium. How many kilograms of sodium chloride are required to produce 2.00 tons

of chlorine?

40.   Silver reacts with nitric acid in the following manner:

$$Ag \; + \; 2\,HNO_3 \longrightarrow NO_2 \; + \; AgNO_3 \; + \; H_2O$$

   (a)   How many grams of HNO_3 are required to react with 27.0 grams of silver?
   (b)   How many grams of NO_2 will be formed?
   (c)   How many grams of $AgNO_3$ will be formed?

41.   How many tons of ammonia will be produced when 1.00 ton of nitrogen gas is reacted with 0.500 ton of hydrogen gas. (Hint: you must determine which of these reactants is the limiting reactant.)

Answers to Check Your Understanding Questions in Chapter 6

1. One hydrogen molecule reacts with one sulfur atom to produce one molecule of hydrogen sulfide molecule. **2.** One calcium atom reacts with two water molecules to form one calcium hydroxide ion group and one molecule of hydrogen. **3.** One copper atom reacts with two silver nitrate ion groups to form two silver atoms and one copper nitrate ion group. **4.** Two water molecules and one carbon atom react to form one molecule of carbon dioxide and two molecules of hydrogen. **5.** One molecule of fluorine reacts with two sodium chloride ion groups to form two sodium fluoride ion groups and one chlorine molecule.

6. $2\,HCl \; + \; Fe \longrightarrow FeCl_2 \; + \; H_2$
7. $N_2 \; + \; 2\,O_2 \longrightarrow 2\,NO_2$
8. $2\,H_3PO_4 \; + \; 3\,Ca(OH)_2 \longrightarrow Ca_3(PO_4)_2 \; + \; 6\,H_2O$
9. $3\,Pb \; + \; Al_2O_3 \longrightarrow 2\,Al \; + \; 3\,PbO$
10. $4\,Al \; + \; 3\,O_2 \longrightarrow 2\,Al_2O_3$
11. $CH_4 \; + \; 2\,O_2 \longrightarrow CO_2 \; + \; 2\,H_2O$
12. $3\,Ni \; + \; 2\,As \longrightarrow Ni_3As_2$
13. $2\,Al \; + \; N_2 \longrightarrow 2\,AlN$
14. $2\,C_2H_2 \; + \; 5\,O_2 \longrightarrow 4\,CO_2 \; + \; 2\,H_2O$
15. $3\,Mg \; + \; Fe_2O_3 \longrightarrow 3\,MgO \; + \; 2\,Fe$
16. $4\,K \; + O_2 \longrightarrow 2\,K_2O$
17. $Sn \; + \; Br_2 \longrightarrow SnBr_2$

18. $3\,Mg\ +\ N_2\ \longrightarrow\ Mg_3N_2$

19. $2\,H_2S\ +\ 3\,O_2\ \longrightarrow\ 2\,SO_2\ +\ 2\,H_2O$

20. $Ca(OH)_2\ +\ Na_2CO_3 \longrightarrow 2\,NaOH\ +\ CaCO_3$

21. a **22.** c **23.** b **24.** b

25. d, $2.80\ \cancel{mol} \times \dfrac{9.01\ g}{1\ \cancel{mol}} = 25.2\ g$ **26.** a, $0.360\ \cancel{mol} \times \dfrac{40.0\ g}{1\ \cancel{mol}} = 14.4\ g$

27. c, $3 \times 10^{22}\ \cancel{atoms} \times \dfrac{1\ \cancel{mol}}{6 \times 10^{23}\ \cancel{atoms}} \times \dfrac{4.00\ g}{1\ \cancel{mol}} = 0.2\ g$

28. d **29.** c, $(2\ x\ 27.0) + 3[12.0 + (3\ x\ 16.0)] = 234$ amu

30. a **31.** (a) 86.9 amu (b) 213 amu (c) 44.0 amu (d) 217 amu
(e) 123 amu (f) 98.1 amu (g) 2.00 amu (h) 103 amu

32.

$$0.500\ L\ \ x\ \frac{170\ \cancel{torr}}{760\ \cancel{torr}}\ \ x\ \frac{273\ \cancel{K}}{298\ \cancel{K}} = 0.102\ L$$

$$.102\ \cancel{L}\ \ x\ \frac{1\ \cancel{mol}}{22.4\ \cancel{L}}\ \ x\ \frac{44.0\cancel{g}}{1\ \cancel{mol}}\ x\ \frac{1000\ mg}{1\ \cancel{g}} = 201\ mg$$

$$P_2 = 170\ \cancel{torr}\ \ x\ \frac{500\ \cancel{mL}}{65\ \cancel{mL}}\ x\ \frac{288\ \cancel{K}}{298\ \cancel{K}} = 1263\ torr$$

33. (a) $3.00\ \cancel{mol} \times \dfrac{86.9\ g}{1\ \cancel{mol}\ MnO_2} = 261\ g$

(b) $2.00 \times 10^{21}\ \cancel{atoms} \times \dfrac{1\ \cancel{mol}}{6.02 \times 10^{23}\ \cancel{atoms}} \times \dfrac{213\ g}{1\ \cancel{mol}} = 0.708\ g$

(c) $0.250\ \cancel{mol} \times \dfrac{44.0\ g}{1\ mol} = 11.0\ g$ (d) $4.50\ \cancel{mol} \times \dfrac{217\ g}{1\ mol} = 976\ g$

34. (a) $2.40 \times 10^{24}\ atoms \times \dfrac{1\ mol}{6.02\ x\ 10^{23}\ atoms} = 4.00\ mol$

(b) $0.482\ \cancel{g} \times \dfrac{1\ mol}{98.1\ \cancel{g}} = 0.00491\ mol$

(c) $1.00\ kg \times \dfrac{1000\ \cancel{g}}{1\ kg} \times \dfrac{1\ mol}{2.00\ \cancel{g}} = 500\ mol$

(d) $4.60\ mg \times \dfrac{1\ \cancel{g}}{1000\ mg} \times \dfrac{1\ mol}{103\ \cancel{g}} = 4.46 \times 10^{-5}\ mol$

35. (a) One molecule of nitrogen reacts with three molecules of hydrogen to produce two molecules of ammonia. (b) One mole of nitrogen reacts with three moles of hydrogen to produce two moles of ammonia. (c) 28.0 grams of nitrogen react with 6.00 grams of hydrogen to produce 34.0 grams of ammonia. **36.** (a) One formula unit of iron(III) chloride reacts with three formula units of sodium hydroxide to form one formula unit of iron(III) hydroxide and three formula units of sodium chloride. (b) One mole of iron(III) chloride reacts with three moles of sodium hydroxide to form one mole of iron(III) hydroxide and three moles of sodium chloride. (c) 162 grams of iron(III) chloride reacts with 120 grams of sodium hydroxide to form 107 grams of iron(III) hydroxide and 176 grams of sodium chloride. **37.** (a) One molecule of fluorine reacts with two formula units of sodium chloride to produce two formula units of sodium fluoride and one molecule of chlorine. (b) One mole of fluorine will react with two moles of sodium chloride to produce two moles of sodium fluoride and one mole of chlorine. (c) 38.0 grams of fluorine will react with 117 grams of sodium chloride to produce 84.0 grams of sodium fluoride and 71.0 grams of chlorine.

38.
$$4\,NH_3 + 3\,O_2 \longrightarrow 2\,N_2 + 6\,H_2O$$

$$51.0\ \text{g}\ NH_3 \times \frac{1\ \text{mol } NH_3}{17.0\ \text{g}} \times \frac{6\ \text{mol } H_2O}{4\ \text{mol } NH_3} \times \frac{18.0\ \text{g}}{1\ \text{mol } H_2O} = 81.0\ \text{g}$$

39.
$$2\,NaCl \longrightarrow 2\,Na + Cl_2$$

$$2.00\ \text{tons} \times \frac{1000\ \text{kg}}{1\ \text{ton}} \times \frac{1\ \text{mol } Cl_2}{0.0710\ \text{kg}} \times \frac{2\ \text{mol } NaCl}{1\ \text{mol } Cl_2} \times \frac{0.0585\ \text{kg}}{1\ \text{mol } NaCl}$$
$$= 3.30 \times 10^3\ \text{kg}$$

40.

(a) $27.0\ \text{g}\ Ag \times \dfrac{1\ \text{mol } Ag}{108\ \text{g}} \times \dfrac{2\ \text{mol } HNO_3}{1\ \text{mol } Ag} \times \dfrac{63.0\ \text{g}}{1\ \text{mol } HNO_3} = 31.5\ \text{g}$

(b) $27.0\ \text{g}\ Ag \times \dfrac{1\ \text{mol } Ag}{108\ \text{g}} \times \dfrac{1\ \text{mol } NO_2}{1\ \text{mol } Ag} \times \dfrac{46.0\ \text{g}}{1\ \text{mol } NO_2} = 11.5\ \text{g}$

(c) $27.0\ \text{g}\ Ag \times \dfrac{1\ \text{mol } Ag}{108\ \text{g}} \times \dfrac{1\ \text{mol } AgNO_3}{1\ \text{mol } Ag} \times \dfrac{170\ \text{g}}{1\ \text{mol } AgNO_3} = 42.5\ \text{g}$

41.
$$N_2 + 3\,H_2 \longrightarrow 2\,NH_3$$

$$1\ \text{ton } N_2 \times \frac{2000\ \text{lb}}{1\ \text{ton}} \times \frac{454\ \text{g}}{1\ \text{lb}} \times \frac{1\ \text{mol } N_2}{28.0\ \text{g}} = 3.24 \times 10^4\ \text{mol Nitrogen}$$

$$0.500 \ \cancel{\text{ton}} \ H_2 \times \frac{2000 \ \cancel{\text{lb}}}{1 \ \cancel{\text{ton}}} \times \frac{454 \ \cancel{g}}{1 \ \cancel{\text{lb}}} \times \frac{1 \ \text{mol} \ H_2}{2.02 \ \cancel{g}} = 22.5 \times 10^4 \ \text{mol Hydrogen}$$

$$3.24 \times 10^4 \ \cancel{\text{mol} \ N_2} \times \frac{3 \ \text{mol} \ H_2}{1 \ \cancel{\text{mol} \ N_2}} = 9.72 \times 10^4 \ \text{mol Hydrogen}$$

Nitrogen is the limiting reagent:

$$3.24 \times 10^4 \ \cancel{\text{mol} \ N_2} \times \frac{2 \ \cancel{\text{mol} \ NH_3}}{1 \ \cancel{\text{mol} \ N_2}} \times \frac{17.0 \ \cancel{g}}{1 \ \cancel{\text{mol} \ NH_3}} \times \frac{1 \ \cancel{\text{lb}}}{454 \ \cancel{g}} \times \frac{1 \ \text{ton}}{2000 \ \cancel{\text{lb}}}$$

$$= 1.21 \ \text{ton} \ NH_3$$

Chapter 7 RADIOACTIVITY AND THE LIVING ORGANISM

In this chapter we focus our attention on the center of the atom—the nucleus—and the changes that can take place within the nucleus to make the atom more stable. We also examine the effects of radiation on living organisms, some methods for measuring this radiation, and the ways in which radiation can be used to benefit humans.

7.1   What is Radioactivity?

Some nuclei with a large ratio of positive protons to neutral neutrons can be unstable, and can give off a particle to form a more stable daughter nucleus. Elements or compounds that give off these particles (or undergo decay) are said to be radioactive. The daughter nucleus formed may also be unstable, and the process of decay will continue until a stable nucleus is formed. Such a sequence of decays is called a decay series or transformation series. Isotopes of an element that give off radiation are called radionuclides or radioisotopes.

Important Terms

radioactivity	transformation series	decay series
radionuclide	daughter nucleus	radioisotope

Alpha Radiation

There are several ways in which a nucleus can decay to become more stable. One way is to give off an alpha particle, made up of two protons and two neutrons (the nucleus of a helium atom). The following is the nuclear equation for the decay of an actinium-225 nucleus, which is an alpha emitter.

$$^{225}_{89}\text{Ac} \longrightarrow {}^{221}_{87}\text{Fr} + {}^{4}_{2}\text{H}$$

(Remember that the sum of the atomic numbers on each side of the arrow must be equal, and the sum of the mass numbers on each side of the arrow must be equal.)

Important Terms

alpha radiation	alpha particle	nuclear equation

Beta Radiation

Beta radiation is composed of streams of electrons, produced within the nucleus and then given off during decay. The daughter nucleus produced in a beta decay will have the same mass number, but a different atomic number than the original nucleus. The following is an example of a beta decay:

$$\ _{28}^{66}\text{Ni} \longrightarrow \ _{29}^{66}\text{Cu} \ + \ _{-1}^{0}\text{e}$$

Important Terms

       beta radiation   beta particle

Gamma Radiation

As discussed in chapter 2, gamma rays are a form of electromagnetic radiation having very short wavelength and very high energy. Gamma rays are often given off in an alpha or beta decay as the daughter nucleus reaches a lower, more stable energy state. Gamma radiation is shown by including the Greek letter gamma (γ) in the nuclear equation. For example, the thorium-234 isotope is a beta and gamma emitter.

$$\ _{90}^{234}\text{Th} \longrightarrow \ _{91}^{234}\text{Pa} \ + \ _{-1}^{0}\text{e} \ + \ \gamma$$

Important Terms

       gamma radiation   gamma rays

Example _____

      Polonium-210 is a beta emitter. Write the shorthand representation of this decay.

S   We need to figure out the other product of this decay and then write the balanced equation for the decay of polonium-210.

T   Use the inside front cover of the text to write the necessary symbols

$$\text{Polonium-210} = \ _{84}^{210}\text{Po}$$

$$\text{Beta particle} = \ _{-1}^{0}\beta$$

Write the equation.

$$\ _{84}^{210}\text{Po} \longrightarrow \ _{N}^{M}\text{X} \ + \ _{-1}^{0}\beta$$

E $210 = M - 0 = 210$

$$84 = N - (-1) = 85$$

P   The symbol for atomic number 85 is At. The balanced equation is:

$$^{210}_{84}Po \longrightarrow {}^{210}_{85}At + {}^{0}_{-1}\beta$$

Check Your Understanding _____

For questions 1 to 5, fill in the blank with the correct word or words.

1.   A substance whose atoms give off particles to become more stable is said to be

      _____.

2.   _____ are nuclei of helium atoms.

3.   Beta radiation consists of streams of _____.

4.   _____ are a form of electromagnetic radiation with very _____ wavelengths.

5.   Gamma rays have _____ charge; alpha particles have a _____ charge;
      and beta particles have a _____ charge.

6.   Without looking back in the text or study guide, write the symbol for each of the
      following:

(a)	proton	(c)	electron	(e)	gamma ray
(b)	neutron	(d)	alpha particle	(f)	beta particle

For questions 7 to 12, fill in the missing symbol.

7. $^{164}_{64}Gd \longrightarrow {}^{144}_{62}Sm + $ _____   10. _____ $\longrightarrow {}^{241}_{95}Am + {}^{4}_{2}He$

8. $^{90}_{38}Sr \longrightarrow {}^{90}_{39}Y + $ _____   11. $^{210}_{84}Po \longrightarrow$ _____ $+ {}^{4}_{2}He$

9. $^{104}_{47}Ag \longrightarrow {}^{0}_{-1}e + $ _____   12. _____ $\longrightarrow {}^{73}_{32}Ge + {}^{0}_{-1}e$

7.2   Half-Life

Each radioactive substance decays at its own characteristic rate. The rate of decay is
measured by the half-life of the radionuclide. The half-life is the amount of time it takes for
one-half of the atoms in a sample to decay. The more unstable the nucleus, the more rapidly
it will decay and the shorter the half-life. The half-lives of radioactive substances provide a
useful tool for establishing the age of human artifacts, once-living organisms, and geologic
periods. Difficulties in solving half-life problems are quite often caused by not accurately
determining the number of half-lives that have elapsed.

Important Terms

        half-life ($t_{1/2}$)   radioisotopic dating

Example

1.   Cesium-137 is a beta emitter. How much of a 5.00 gram sample would remain after 90.6 years?

S   We want to know: ^{137}Cs = (?) g after 90.6 years.

T   First we must decide how many half-lives will occur in 90.6 years. From Table 7.2 we learn that cesium-137 has a half-life of 30.2 years. We can use this fact to determine the number of half-lives. Then we can calculate the amount left after n half-lives.

E

$$90.6 \; \cancel{\text{years}} \; \text{X} \; \frac{1 \; \text{half-life}}{30.2 \; \cancel{\text{years}}} = 3 \; \text{half-lives}$$

$$^{137}Cs \; \text{remaining} \; = \; 5.00 \; \text{g} \; \text{x} \; \left(\frac{1}{2}\right)^3$$

$$= \frac{5.00 \; \text{g}}{8} \; = 0.625 \; \text{g}$$

P   The answer should have three significant figures and is 0.625 g.

2.   Strontium-90 is a beta emitter with a half-life of 29.0 years. How long will it take for a 12.0 g sample of $^{90}_{38}Sr$ to be reduced to 1.5 g?

S   The problem asks us to find out how long it takes for the sample to be reduced from 12.0 g to 1.5 g.

T   First we must decide what fraction of the sample remains. From this we can determine how many half-lives have elapsed. Then we can multiply the number of half-lives by 29.0 years to determine the amount of time that has elapsed.

E

$$\frac{1.5}{12.0} = \frac{1}{8} = \frac{1}{2^3} \qquad \text{Three half-lives will have elapsed.}$$

$$3 \; \cancel{\text{half-lives}} \; \text{x} \; \frac{29.0 \; \text{yr}}{\cancel{\text{half-life}}} = 87 \; \text{years}$$

P   It will take 87 years for the sample to decay from 12.0 g to 1.25 g

Check Your Understanding _____

13.   The half-life of fermium-253 is 4.5 days. How many grams of a 2.0-gram sample
       would remain after 13.5 days?
14.   Iodine-128 is used in medical diagnosis and has a half-life of 25 minutes. What
       fraction of the original sample would remain after 2 1/2 hours?

7.3   Nuclear Fission
To meet the energy needs of our society, we are harnessing the energy produced by nuclear
reactions. The nuclei of certain isotopes can be broken apart to form smaller, more stable
nuclei. This process, called fission, occurs when one of these isotopes is struck by a neutron.
The isotope then breaks apart to form two smaller nuclei, two or three neutrons, and a
tremendous amount of energy. The neutrons so produced can react with other fissionable
nuclei to cause further reactions. If a critical mass of fuel nuclei is present, this process can
lead to a chain reaction.

Important Terms

       fission   chain reaction   critical mass

7.4   Nuclear Wastes
The radioactive wastes produced in the fission process have long half-lives and must be
stored safely for a long time. These wastes, which are found in spent fuel from nuclear
power plants and from nuclear-related defense activities, are classified as high-level wastes.
Other sources of radioactive wastes are mill tailings, waste from medical and commercial
processes, and transuranic waste. High-level wastes are accumulating in temporary storage
facilities around the country. The U.S. Department of Energy is responsible for establishing
a permanent storage facility for high-level wastes, but location of this site is still being
debated.

7.5   Nuclear Fusion
Small nuclei can also react by combining with one another to form heavier, more stable
nuclei. This process, called nuclear fusion, yields a tremendous amount of energy. Nuclear
fusion reactions generate the energy given off by our sun. If we could develop the
technology necessary to harness nuclear fusion reactions, this process would have great
advantages over nuclear fission. Delay in developing this technology is caused by the fact
that nuclear fusion reactions require temperatures of one hundred million degrees Celsius or
more to occur.

Important Terms

nuclear fusion   deuterium   tritium

Check Your Understanding _____

For questions 15 to 19, choose the best answer or answers from the choices given.

15.   This reaction involves the joining of smaller nuclei to form a larger, more stable nucleus.
        (a)   nuclear fusion   (c)   nuclear transmutation
        (b)   nuclear fission   (d)   beta decay

16.   This reaction involves the splitting apart of a large nucleus into two smaller, more stable nuclei.
        (a)   nuclear fusion   (c)   nuclear transmutation
        (b)   nuclear fission   (d)   beta decay

17.   A requirement for nuclear fusion to occur is
        (a)   a critical mass of uranium-235
        (b)   a moderator
        (c)   control rods
        (d)   very high temperatures

18.   A nuclear chain reaction can occur only when
        (a)   the temperature is very high
        (b)   a moderator is present.
        (c)   a critical mass of fuel nuclei are present
        (d)   a heat transfer fluid is present

19.   Which of the following are advantages of nuclear fusion over nuclear fission?
        (a)   It is more efficient.
        (b)   It doesn't require high temperatures.
        (c)   The fuel is abundant.
        (d)   There is little radioactive waste.
        (e)   There is little chance of the reactor core melting.
        (f)   All of the above.

7.6   Ionizing Radiation

Alpha, beta and gamma radiation each have different penetrating power and interactions with living tissue. Because of its large size, an alpha particle has very little penetrating power. Beta particles are much smaller and have more penetrating power. Gamma radiation has high penetrating power.

Radiation interacts with living cells to produce highly reactive particles. These particles may be charged particles, called ions, or high-energy, uncharged particles called free radicals. Both ions and free radicals produce harmful changes in living cells. Cells have efficient repair mechanisms to protect against damage to DNA, and contain molecules such as vitamin C and A that quickly neutralize any ions or free radicals that form. Cells with changed or altered DNA are called mutant cells. If such mutant cells divide and grow in an uncontrolled fashion, they are called cancerous or malignant.

Important Terms

ionizing radiation ion   free radical
malignant cell   mutant cell cancerous cell

7.7 Radiation Dosage

This section discusses various units used to measure ionizing radiation, or the effects of the radiation on living tissues. Table 7.3 lists several units used to measure ionizing radiation. The curie measures the rate of radioactive emissions from a radioactive source. It does not tell us anything about the type of radiation produced or the effect of this radiation. The rad measures the amount of energy released per gram of tissue. The rem indicates the absorbed dosage of radiation independent of the type of radiation used. The biological effects of one rem of each type of radiation are the same.

Important Terms

curie (Ci)   rem   rad

7.8 Background Radiation

Each of us is exposed to a small amount of ionizing radiation each day from natural sources. This radiation is called background radiation, and is produced by naturally occurring radionuclides that may be in the soil, food, water, or air. The average person receives only a small dose of radiation (about 400 mrems) per year.

Important Term

background radiation

7.9 Protection Against Radiation Exposure

To protect yourself from the damaging effects of ionizing radiation, you should minimize your exposure to it, remain as far away from the source of radiation as possible, and use proper shielding.

Important Terms

shielding   inverse square law

Check Your Understanding ⎯⎯⎯⎯⎯⎯⎯⎯⎯⎯⎯⎯⎯⎯⎯⎯⎯⎯⎯⎯⎯⎯⎯⎯

20. Radiation causes harmful effects in living tissues by producing highly reactive particles called _____ and _____.

21. Match each of the following definitions with the correct unit or units for measuring radiation. (Use Table 7.3)

(a)	Produces the same biological effect as one rad of therapeutic X rays	A. becquerel
		B. rem
(b)	Measures only the frequency of disintegrations	C. rad
		D. curie
(c)	Measures a specific amount of energy released per gram of irradiated tissue	E. gray
		F. sievert
(d)	The SI unit that equals 100 rem.	

For questions 22 to 27, indicate whether the statement is true (T) or false (F).

22. The average person is exposed to more ionizing radiation from the natural environment than from any other source.

23. The biological effects of one rad of various types of radiation will be the same.

24. The cell has mechanisms to repair the damage caused by ionizing radiation.

25. Cells with altered DNA are called mutant cells.

26. You would receive a more intense dose of radiation from a sample of cobalt-60 standing 9 feet from the sample than 3 feet from it.

27. Alpha particles are highly penetration ionizing radiation.

7.10   Medical Diagnosis

Radioactive chemicals called tracers are increasingly useful in medical diagnosis. Both natural and artificially produced radionuclides are used. Compounds containing radioactive elements are synthesized and used to examine a specific area of the body (such as the thyroid or brain), or a specific function in the body (such as blood flow or the movement of bile). With the development of highly sensitive measuring devices, diagnostic pictures can be produced using radionuclides that expose the patient to much lower dosages. A CT scanner uses X rays to produce detailed pictures of cross-sections of the human body, eliminating the need for many invasive diagnostic tests and exploratory surgeries. PET scans record gamma radiation produced when positrons, given off by a radioactive tracer, interact with electrons in the body. These scans produce important information about brain function and the development of disease.

tracer   gamma camera   CT scanner   PET scan

7.11   Radionuclides Used in Diagnosis

Technetium-99^m is a gamma emitter with a half-life of six hours. It is used extensively in the study of the liver, spleen, heart and bone marrow. Because of its desirable properties, technetium-99^m is replacing most other isotopes in common diagnostic procedures.

Important Terms

metastable   metastasize

7.12   Radiation Therapy

Ionizing radiation can be used to treat cancerous tissue because cancer cells divide more rapidly than normal cells and are more sensitive to the effects of ionizing radiation. High-intensity radiation from an X-ray machine, a cobalt-60 source, or a particle accelerator is used to treat cancerous tissue that cannot be removed by surgery. To treat certain other cancers, radionuclides are inserted into the tumor with a needle. Scientists have also synthesized radionuclide-containing chemicals that concentrate in the region or tissue that is to be treated.

Check Your Understanding ⎯⎯⎯⎯⎯⎯⎯⎯⎯⎯⎯⎯⎯⎯⎯⎯⎯⎯⎯⎯⎯⎯⎯

For questions 28 to 38, choose the best answer or answers.
28.   This is effective in locating and diagnosing brain tumors.
   (a)   gamma camera scan   (c)   PET scan
   (b)   CT scan   (d)   MRI scan
29.   This isotope is used to treat thyroid cancer.
   (a)   iodine-131   (c)   strontium-90
   (b)   radium-226   (d)   phosphorus-32
30.   This isotope acts like potassium and concentrates in undamaged heart muscle.
   (a)   iodine-131   (c)   thallium-201
   (b)   radium-226   (d)   phosphorus-32
31.   Instrument used to produce a picture showing the location of a radioactive tracer.
   (a)   gamma camera   (c)   PET scanner
   (b)   CT scanner   (d)   MRI scanner
32.   Instrument used to follow the metabolic activity of the brain.
   (a)   gamma camera   (c)   PET scanner
   (b)   CT scanner   (d)   MRI scanner

33. Instrument that produces detailed cross-sectional images of the body.
    (a) gamma camera (c) PET scanner
    (b) CT scanner (d) MRI scanner
34. This isotope is inserted by needle to slow tumor growth.
    (a) thallium-201 (c) technetium-99^m
    (b) radium-226 (d) gold-198
35. When cancer cells spread from the original tumor, they have
    (a) translated (c) metastasized
    (b) transformed (d) replicated
36. This isotope is in an energy state that is higher than normal.
    (a) thallium-201 (c) technetium-99^m
    (b) radium-226 (d) gold-198
37. This isotope is used to detect if cancer cells from a tumor have spread to the bones.
    (a) phosphorus-32 (c) strontium-90
    (b) technetium-99^m (d) radium-226
38. Chemicals that contain radioactive atoms and that can be used to monitor living functions are called
    (a) metastable (c) tracers
    (b) isotopes (d) radionuclides

For questions 39 to 44, fill in the blank with the correct word or words.
39. Three advantages of technetium-99^m over other radioactive tracers are _____, _____, and _____.
40. A _____ analyzes the chemical composition of a tissue to show early stages of a disease.
41. A CT scanner forms its image using _____.
42. The isotope technetium-99^m is in the _____ state and releases _____ to become more stable.
43. Cancerous tissue that cannot be removed by surgery might be treated by _____, _____, and _____.
44. Cells that divide _____ (slowly, rapidly) are the most sensitive to ionizing radiation.

Select the best option for each part of question 45
45. A radiologist selecting a radioisotope for use in diagnosis would try to find one that
    (a) Has a very long half-life, or has a short half-life
    (b) Is quickly eliminated, or is retained by the body
    (c) Will give results using 10 mg, or will give results using 50 mg

Answers to Check Your Understanding Questions in Chapter 7

1. radioactive 2. alpha particles 3. electrons 4. gamma rays, short 5. no, positive, negative

6. (a) $_1^1p$ (b) $_0^1n$ (c) $_{-1}^0e$ (d) $_2^4He$ (e) γ (f) $_{-1}^0e$

7. $_2^4He$ 8. $_{-1}^0e$ 9. $_{48}^{104}Cd$ 10. $_{97}^{245}Bk$ 11. $_{82}^{206}Pb$ 12. $_{31}^{73}Ga$

13. $13.5 \text{ days} \times \dfrac{1 \text{ half-life}}{4.5 \text{ days}} = 3$ half-lives

    grams remaining $= 2.0 \text{ g} \times (\tfrac{1}{2})^3 = 0.25$ g

14. $2.5 \text{ hr} \times \dfrac{60 \text{ min}}{1 \text{ hr}} \times \dfrac{1 \text{ half-life}}{25 \text{ min}} = 6$ half-lives

    $(\tfrac{1}{2})^6$ or 1/64 of the sample will remain

15. a 16. b 17. d 18. c 19. a, c, d, e 20. ions, free radicals 21. (a) B (b) A, D (c) C, E (d) F
22. T 23. F 24. T 25. T 26. F 27. F 28. b 29. a 30. c 31. a, c 32. c 33. b, d 34. d 35. c
36. c 37. b 38. c 39. short half-life, gives off only gamma radiation, and the energy of the
gamma radiation can be easily detected 40. PET scan 41. X rays 42. metastable, gamma
radiation 43. irradiation with X rays, insertions of a radionuclide into the tumor, or treatment
with a chemical containing a radionuclide 44. rapidly 45. (a) short half-life (b) quickly
eliminated (c) will give results using 10 mg

Chapter 8   REACTION RATES AND CHEMICAL EQUILIBRIUM

Why do chemical reactions proceed at different rates? Why do some reactions give off heat, while others cannot occur unless heat is added? What factors can change the rate of a chemical reaction? Why do some chemical reactions proceed for a while and then seem to stop? These are a few of the questions about chemical reactions answered in this chapter. To discuss reaction rates and chemical equilibrium, we must introduce some new vocabulary. Be sure that you understand the important terms in each section.

8.1   Activation Energy

For a chemical reaction to occur, the reactant particles must collide with enough energy to overcome the forces of repulsion between the electrons surrounding each particle. In addition, the particles must collide with enough energy to break old bonds so that new bonds can be formed. The activation energy is the amount of energy necessary for a collision between reactant particles to result in the formation of products. Each reaction has a characteristic activation energy. The higher the activation energy, the greater the energy the reactant particles must possess to have successful collisions and form products.

Important Terms

       activation energy   activated complex   transition state

8.2   Exothermic and Endothermic Reactions

Chemical reactions can be classified as exothermic (those that give off energy) or endothermic (those that require energy to occur). The heat of reaction, ΔH, is the amount of energy given off or required in a reaction. The value of the heat of reaction is equal to the difference between the potential energy of the products and the potential energy of the reactants. A potential energy diagram shows the relationship between the potential energy of the reactants and products. Study carefully the diagrams for an exothermic and an endothermic reaction shown in Figure 8.4 in the text. From these diagrams you can see that the value of ΔH will be negative if the reaction is exothermic, and positive if the reaction is endothermic.

Important Terms

       endothermic reaction   exothermic reaction
       potential energy diagram   heat of reaction (ΔH)

For questions 1 to 5, answer true (T) or false (F).

1. The heat of reaction is the minimum energy required in a collision between reactant particles for a reaction to take place.
2. An endothermic reaction gives off energy.
3. The products will have less potential energy than the reactants in an exothermic reaction.
4. The activation energy of a reaction is the difference between the potential energy of the products and the potential energy of the reactants.
5. If the activation energy barrier is high, most collisions between reactant molecules at room temperature will not be successful.

6. Identify each of the following reactions as exothermic or endothermic.

    (a) $2 H_2 + O_2 \longrightarrow 2 H_2O + 115.6 \text{ kcal}$
    (b) $H_2 + I_2 + 12.4 \text{ kcal} \longrightarrow 2 HI$
    (c) $SO_2 + 71 \text{ kcal} \longrightarrow S + O_2$
    (d) $2 C + 3 H_2 \longrightarrow C_2H_6 + 20.2 \text{ kcal}$

7. The following is the potential energy curve for the reaction

$$N_2 + 2O_2 \longrightarrow 2NO_2$$

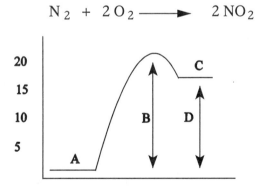

    (a) What letter represents $2NO_2$?
    (b) What letter represents $N_2 + 2O_2$?
    (c) What letter represents the activation energy?
    (d) What letter represents the heat of reaction?
    (e) Is the reaction endothermic or exothermic?
    (f) What is the numerical value of ΔH?

8. Draw the potential energy diagrams for two exothermic reactions having the same heats of reaction, but one of which is very fast at room temperature and one very slow.

8.3   Factors Affecting Reaction Rates

There are many factors that influence the rate at which a chemical reaction will occur. The first is the nature of the reactants themselves. This includes the stability of the molecules, the type of bonding, the size and shape of the molecules, and the physical states of the reactants (whether they are solids, liquids, or gases).

The second factor that affects the rate of a reaction is the concentration of the reactants. The higher the concentration of reactants, the larger the number of reactant particles in a given volume, and the higher the probability that collisions will occur. If the number of collisions is increased, it is more likely that there will be successful collisions.

Because particles must collide in order for a reaction to occur, only particles of a solid that are on the surface can undergo a reaction. Increasing the surface area of a solid will increase the number of particles on the surface. This increases the number of particles available to react (the concentration of reactant) and, therefore, increases the reaction rate.

Increasing the temperature of the reactants increases the kinetic energy possessed by the reactant particles. This increases the reaction rate for two reasons: it increases the number of collisions (because the particles are moving faster) and the energy of the collisions. Similarly, decreasing the temperature of a reaction will decrease the rate of the reaction.

A catalyst is a substance that increases the rate of a chemical reaction by lowering the activation energy barrier. Because the catalyst is not consumed in the reaction, it can be used over and over. This means it can be present in very small amounts. Catalysts are widely used in the chemical industry to increase yields and lower the cost of producing chemicals.

The reactions that occur in our bodies take place under very mild conditions of temperature and pressure and at very low reactant concentrations. Most of the reactions that occur in our bodies would proceed very slowly at body temperature without catalysts called enzymes.

Important Terms

catalyst   reaction rate   enzyme

Check Your Understanding _____

9.   What effect will each change listed below have on the rate of the following reaction?

$$CaCO_3 + 2\,HCl \longrightarrow CaCl_2 + CO_2 + H_2$$
(marble)

   (a)   Grind up the marble into fine particles.
   (b)   Use a lower concentration of hydrochloric acid.
   (c)   Place the reaction container in an ice bath.
   (d)   Add a catalyst.

10.   Draw the potential energy diagram for an endothermic reaction. Using a dotted line, show what the potential energy diagram would look like if a catalyst were added to the reaction.

11.   In what way can poisons affect the rate of a chemical reaction in living organisms?

8.4   What is Chemical Equilibrium?

Chemical equilibrium is a dynamic state on the molecular level that seems static on the macroscopic level, the level that we can see. It is a dynamic state in which the rate of the forward reaction equals the rate of the reverse reaction. For a system in chemical equilibrium, there is no net change in the number of reactant or product particles, even though the reactants are constantly forming products and the products are constantly forming reactants. For a reaction to reach equilibrium, two conditions must be met: a uniform temperature must be maintained and, after the reaction has started, no substances can be added or removed from the system. Remember the key phrase: the only thing that is equal at equilibrium is the rate of the forward and reverse reactions.

When water is placed in an open container at 25°C, it will evaporate until it is gone. But if water is placed in a closed flask at 25°C, from which all the gas has been evacuated, and if the pressure in the flask is then monitored, the pressure will first increase until it reaches 23.8 mm Hg and then will remain constant as long as the temperature remains constant. On the molecular level, the water molecules begin to evaporate as soon as the water is placed in the evacuated flask.

$$H_2O_{(l)} \longrightarrow H_2O_{(g)}$$

As the number of gas molecules increases, they begin to collide and condense to form liquid water.

$$H_2O_{(l)} \rightleftharpoons H_2O_{(g)}$$

The more gas molecules, the faster the rate of condensation until the rate of condensation equals the rate of evaporation. The system is then in equilibrium. This occurs when the pressure of the water vapor equals 23.8 mm Hg.

$$H_2O_{(l)} \rightleftharpoons H_2O_{(g)}$$

The number of water molecules entering the gas phase each second will equal the number of water molecules entering the liquid phase each second. Thus, equilibrium is a dynamic state in which changes occur at the same rate.

Important Terms

equilibrium   dynamic state

Check Your Understanding

12.   A chunk of iodine is placed in a water-alcohol mixture, and a reddish color quickly appears around the solid. The mixture is stirred and the color of the liquid initially deepens, and then no further change in the color of the liquid or in the mass of the iodine remaining on the bottom of the container can be detected. The temperature is held constant.

(a)   Has an equilibrium been established in this situation? Give reasons for your answer.

(b)   State in words what has happened on the molecular level.

(c)   State in equation form what has happened on the molecular level.

8.5 Altering the Equilibrium

A system at equilibrium is one in which the forward and reverse reactions are occurring at the same rate. This chemical equilibrium can be disrupted by any factor that changes the rate of either the forward or reverse reaction without changing the rate of the other. Increasing the concentration of a reactant increases the rate of the forward reaction, while increasing the concentration of a product increases the rate of the reverse reaction. Similarly, decreasing the concentration of a reactant decreases the rate of the forward reaction and decreasing the concentration of a product decreases the rate of the reverse reaction. Changing the temperature of the system will favor one reaction over the other. If the temperature is increased, the endothermic reaction is favored. If the temperature is decreased, the exothermic reaction if favored. Adding a catalyst does not disrupt an equilibrium because a catalyst increases the rate of both the forward and the reverse reactions by equal amounts.

8.6 Le Chatelier's Principle

Le Chatelier's principle allows us to predict the effect of such changes on the equilibrium concentrations of reactants and products. Le Chatelier's principle states that if a stress is applied to a system in equilibrium, the system will shift in a direction that will remove the stress. The stress could be a change in concentration of one or more reactants or products, a change in temperature or a change in pressure (when gaseous reactants or products are involved). It is important to remember that when stress is applied to a system at equilibrium, it will come to a new equilibrium state, it does not return to the original state.

Important Term

Le Chatelier's principle

Example _____

What is the effect of each of the following on the equilibrium concentration of HI in the following system?

$$2\,HI \; \rightleftharpoons \; H_2 + I_2 + 12.4\,kcal$$

(a) Adding I_2

(b)   Removing H_2

(c)   Increasing the pressure

(d)   Increasing the temperature

(e)   Adding a catalyst

To answer this question, we will use the *T (Think it Through)* part of our *STEP* method.

T   (a)   Increasing the concentration of I_2 will increase the rate of the reverse reaction, forming more HI. With more HI molecules present, the rate of the forward reaction will then increase until it again equals the rate of the reverse reaction, and a new equilibrium will be established. Based on Le Chatelier's principle we know that the system will shift to the left to relieve the stress caused by adding a product. At the new equilibrium, there will be more HI than at the initial equilibrium.

T   (b)   Removing some H_2 will slow the rate of the reverse reaction. The forward reaction will continue, but it will now be at a greater rate than the slowed down reverse reaction. When enough reactant molecules are used up, the forward reaction will slow down until it is at the same rate as the reverse reaction. Based on Le Chatelier's principle we know that removal of a product will shift the equilibrium to the right. There will be less HI.

T   (c)   The stress in this case is increased pressure. How can the system shift to remove the pressure? The forward reaction produces one molecule of H_2 and one molecule of I_2. H_2 is a gas, but I_2 is a solid. The reverse reaction produces two molecules of HI. Since two molecules will exert twice as much pressure as one, the forward reaction forms a system of lower pressure. Therefore, to remove the stress the system will shift to the right, producing H_2 and I_2. A new equilibrium will be established with more product molecules and fewer reactant molecules.

T   (d)   Increasing the temperature will increase the rate of both reactions, but the rate of the endothermic reaction (the reaction that removes heat) will be favored, forming more molecules of HI. The stress on the system was the increase in temperature, and the system will move in a direction that will remove the stress. The endothermic reaction removes energy and, therefore, is favored when the temperature is increased. The system shifts to the left.

T   (e)   A catalyst speeds up both the forward and reverse reactions the same amount, so it does not affect the position of equilibrium. There will be no shift in this equilibrium. If a system is not already at equilibrium, a catalyst will cause it to come to equilibrium state faster.

Check Your Understanding ──────────────────────────────

Use the following equation to answer questions 13 to 16:

$$2\,NO + O_2 \ \rightleftharpoons\ 2\,NO_2 + 27\,kcal$$

What is the effect of each of the following on (a) the equilibrium concentration of O_2, and (b) the rate of formation of NO_2?
13.   Increasing the concentration of NO (while the temperature remains constant).
14.   Increasing the pressure (with constant temperature)
15.   Increasing the temperature.
16.   Adding a catalyst.

Use the following information to answer questions 17 to 21.

Suppose we have a flask containing dinitrogen tetroxide and nitrogen dioxide gases at equilibrium at 25°C.

The equation for the reaction is

$$N_2O_4\,(g) + 13.9\,kcal \ \rightleftharpoons\ 2\,NO_{2\,(g)}$$
colorless   reddish-brown

17.   What effect would adding more N_2O_4 to the flask have on the equilibrium?
18.   What would be the effect of removing some NO_2?
19.   What would be the effect of doubling the pressure by decreasing the volume (with the

temperature held constant)?

20.   What is the effect on the equilibrium of placing the flask in a water bath at 100°C?

21.   Can you predict what would happen to the color of the flask if it were placed in ice water?

Answers to Check Your Understanding Questions in Chapter 8

1. F **2.** F **3.** T **4.** F **5.** T **6.** (a) exothermic (b) endothermic (c) endothermic (d) exothermic
7. (a) C (b) A (c) B (d) D (e) endothermic (f) $\Delta H = (+)$ 16 kcal/mol
8.

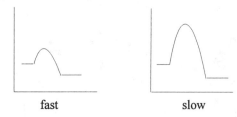

fast   slow

9. (a) increase (b) decrease (c) decrease (d) increase
10.

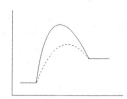

11. They destroy enzymes, thereby decreasing the rate of the reactions in the body.
12. (a) Yes. The temperature is constant and no net change is observed in the system. (b) The solid iodine began to dissolve. As the dissolved molecules began to increase, some combined to reform the solid. A point was reached when the rate of dissolving equaled the rate of reforming the solid, and an equilibrium was established.

(c)   Initially I_2 (s) $\longrightarrow$ I_2 (aq)
       Then I_2 (s) $\rightleftharpoons$ I_2 (aq)
       Finally I_2 (s) $\rightleftharpoons$ I_2 (aq)

13. (a) decrease (b) increase **14.** (a) decrease (b) increase **15.** (a) increase (b) increase **16.** (a) no change (b) increase [Remember a catalyst will raise the rate of both the forward and the reverse reactions] **17.** shift to right **18.** shift to right **19.** shift to left **20.** shift to right **21.** it would turn colorless.

Chapter 9   WATER, SOLUTIONS, AND COLLOIDS

Water is the dissolving fluid in all living organisms. An understanding of water and its properties, and of the properties of substances dissolved in water, is critical to our understanding of how living organisms function. This chapter discusses the properties of water, solutions, and colloids.

9.1   Molecular Shape of Water

Water has many properties that make it unique among chemicals and essential to life processes. The water molecule is polar, and it is an excellent solvent for ionic compounds and polar covalent compounds. Hydrogen bonds can form between water molecules or between water and other molecules. It is the polarity of the water molecule and the possibility of hydrogen bonding that make water such an excellent solvent.

Important Terms

      solvent   universal solvent   polar molecule   hydrogen bonding

9.2   Properties of Water

The polarity of the water molecule and the hydrogen bonds it can form are responsible for several unique properties of water.

- Water is an excellent solvent for ionic and polar compounds.
- The strong attraction between water molecules accounts for its high melting and boiling points when compared to other compounds with comparable molecular weights.
- Because of hydrogen bonding between water molecules in ice, ice is less dense than liquid water at $0°C$. This is why ice floats on water.
- Water molecules at the surface are very strongly attracted to other water molecules, and not to nonpolar air molecules. This gives water a high surface tension.
- Water has a high heat of vaporization; that is, it takes a lot of energy to convert water from the liquid to the gaseous phase. This energy is released when water condenses.
- Water has a high heat of fusion; that is, water gives off a lot of heat when it goes from the liquid to the solid phase. When water goes from the solid to the liquid phase, a lot of heat must be added.
- Water has a high specific heat when compared to other liquids. This means that water can absorb heat without great changes in its temperature.

Important Terms

surface tension   heat of vaporization   heat of fusion
specific heat   surfactant   density

Check Your Understanding _____

For questions 1 to 13, answer true (T) or false (F).
1. Humans can survive several weeks without water.
2. In the human body, water is found within the cells, around the cells, and in the blood plasma.
3. A water molecule is nonpolar, although it contains two polar covalent bonds.
4. Hydrogen bonds can form between a hydrogen on one water molecule and a hydrogen on another water molecule.
5. Methane (CH_4) will dissolve in water.
6. The high melting point of water results from interactions between the water molecules.
7. The density of a substance is a measure of the mass of that substance per unit of volume.
8. Ice is more dense than water.
9. Because of its high surface tension, water will tend to spread out on a nonpolar surface and to form beads on a polar one.
10. Surfactants are substances that act to reduce the surface tension of water.
11. It takes 1.6 kilocalories to melt 20 grams of water.
12. A great deal of heat must be added to water to melt it or to vaporize it.
13. The higher the specific heat of a substance, the greater the change in its temperature when it absorbs a given amount of heat energy.

9.3   Three Important Mixtures
The major difference between suspensions, solutions and colloids is the size of the dissolved particles (See Table 9.2).

9.4   Suspensions
Suspensions are heterogeneous mixtures whose particles will settle out in time and can be separated from the liquid with filter paper or a centrifuge. Suspensions are often opaque and may be translucent. Whole blood is an example of a suspension.

Important Terms

suspension   heterogeneous

9.5   Colloids

The particles that form colloids (or colloidal dispersions) are larger than the particles that form solutions. When placed in a solvent, they never truly dissolve; they are found dispersed in the solvent, but are not heavy enough to settle out. Particles larger than colloids (those forming suspensions) will settle out of solution. Milk is a good example of a colloid. The white appearance of milk is caused by millions of tiny colloidal-size globules of fat suspended in the liquid. Table 9.3 in the text lists examples of the eight classes of colloids.

Colloids exhibit several distinctive properties. Brownian movement is the random movement of colloidal particles caused by the bombardment of these particles by solvent molecules. Colloids can be distinguished from true solutions by the Tyndall effect. In addition, the large surface area of colloidal particles gives them the ability to adsorb large amounts of other substances on their surfaces.

Important Terms

colloidal dispersion	colloid	Brownian movement
Tyndall effect	adsorption	

Check Your Understanding —————————————————————————

For questions 14 to 18, choose the correct answer.

14.   Substances that form solutions when placed in water are
   (a)   colloids   (c)   colloidal dispersions
   (b)   suspensions   (d)   crystalloids

15.   Heterogeneous mixtures containing particles suspended in a liquid that can be separated from the liquid using filter paper are called
   (a)   colloids   (c)   colloidal dispersions
   (b)   suspensions   (d)   crystalloids

16.   If water contains a dispersion of particles larger than ions or molecules, but not large enough to settle out, the mixture is called a
   (a)   precipitate   (c)   colloidal dispersion
   (b)   suspension   (d)   crystalloid

17.   Which of the following is not a property of a colloid?
   (a)   Brownian movement   (c)   can be separated using filter paper
   (b)   adsorption   (d)   the Tyndall effect

18.   You might use this to remove a colored impurity from a solution you are preparing in the laboratory.
   (a)   filter paper   (c)   centrifuge
   (b)   powdered charcoal   (d)   Brownian movement

9.6 Solutions

A solution is a homogeneous mixture whose particles are of atomic or molecular size. The solute is the substance that is being dissolved, and the solvent is the substance that does the dissolving. An aqueous solution is one in which water is the solvent.

Ionic solids will dissolve in water only when the attraction between the ions and the water molecules is greater than the attraction among the ions in the crystal lattice. A hydrated ion is one surrounded by water molecules. Highly polar covalent substances are soluble in water because of polar-polar interactions and the hydrogen bonding that can form between these molecules and water molecules.

Important Terms

solution   solute   solvent
aqueous solution   hydrated ion

9.7 Electrolytes and Nonelectrolytes

Solutes, the particles dissolved in a solution, can be either nonelectrolytes or electrolytes. Nonelectrolytes are uncharged solute particles whose aqueous solution will not conduct electricity. Sugar water is a solution of a nonelectrolyte. It will not conduct electric current. Electrolytes are charged solute particles whose aqueous solutions will conduct electricity. Salt is an electrolyte. Salt water will conduct electric current. Electrolytes can occur in solution in two ways: through the breaking apart of an ionic crystal, or the ionization of a polar covalent compound. The charged solute particles formed are anions (negative particles) and cations (positive particles). A strong electrolyte is one that completely ionizes or breaks apart in solution; a weak electrolyte only partially ionizes or breaks apart.

Important Terms

electrolyte   nonelectrolyte   cation   anion

Check Your Understanding ──────────────────────────────────────

19. Predict whether each of the following are electrolytes (E) or nonelectrolytes (N):
     (a) KCl   (b) $CaSO_4$ (c) HBr   (d) CH_3CH_2OH (alcohol)
20. Indicate which of the following are cations and which are anions.
     (a) Li^+ (b) SO_4^{2-} (c) I^- (d) NH_4^+

9.8 Factors Affecting the Solubility of a Solute

The solubility of a solute indicates the amount of solute that will dissolve in a solvent.

Several factors affect the solubility of a solute in a particular solvent: the nature of the solute and of the solvent, the temperature, and the pressure (if the solute is a gas.) The general rule is "like dissolves like." Polar solutes dissolve in polar solvents and nonpolar solutes dissolve in nonpolar solvents.

Important Term

solubility

9.9 Solubility of Ionic Solids

Ionic solids vary in their solubility. The general rules found in the text allow us to predict whether a particular compound is water soluble. They also allow us to predict if a precipitate will form when two solutions of soluble ionic solids are mixed.

Important Terms

net-ionic equation   precipitate

Example _____

Will a precipitate form when solutions of lead nitrate and sodium sulfide are mixed? If a precipitate forms, write the net ionic equation for the reaction.

S   The question asks us to decide whether an insoluble substance is formed when these two salts are mixed and then, if a precipitate forms, to write the net ionic equation for its formation.

T   A solution of lead nitrate, $Pb(NO_3)_2$, contains lead ions (Pb^{2+}) and nitrate ions (NO_3^-). A solution of sodium sulfide, Na_2S, contains sodium ions (Na^+) and sulfide ions (S^{2-}). The two new substances that could form from a mixture of these two solutions are $NaNO_3$ and PbS. Looking at the table in Section 9.9 in the text, we see that $NaNO_3$ is soluble but PbS is not. A precipitate of PbS will form.

E   The complete equation for this reaction is

$$Pb^{2+}_{(aq)} + 2\,NO_3^-_{(aq)} + 2\,Na^+_{(aq)} + S^{2-}_{(aq)} \longrightarrow PbS_{(s)} + 2\,Na^+_{(aq)} + 2\,NO_3^-_{(aq)}$$

P   To obtain the net-ionic equation, you cancel the ions not involved in the reaction.

$$Pb^{2+}_{(aq)} + S^{2-}_{(aq)} \longrightarrow PbS_{(s)}$$

116

For questions 21 to 24, (a) predict whether a precipitate will form when aqueous solutions containing the two compounds are mixed, and (b) write the net-ionic equation when a reaction does occur.

21.   Potassium sulfide and silver nitrate
22.   Ammonium chloride and sodium acetate
23.   Zinc bromide and sodium carbonate
24.   Barium nitrate and lithium hydroxide

For questions 25 to 32 select the best answer.

25.   Particles in a solution can be separated from colloidal particles by using a membrane because the particles in a solution:

(a)	are smaller	(c)	do not form ions
(b)	are ionic	(d)	are larger

26.   When sodium chloride is dissolved in water, the water is called the:

(a)	solution	(c)	solvent
(b)	solute	(d)	colloid

27.   Which of the following is not an electrolyte

(a)	NaBr	(c)	HCl
(b)	KBr	(d)	CCl_4

28.   Which of the following is a characteristic of a solution?

(a)	settle on standing	(c)	are colorless
(b)	pass through filter paper	(d)	do not pass through membranes

29.   Mayonnaise is:

(a)	a solution	(c)	a suspension
(b)	an emulsion	(d)	an aerosol

30.   Which of the following is not a colloidal dispersion?

(a)	milk of magnesia	(c)	clouds
(b)	milk	(d)	marshmallows

31.   In which of the following are the particles arranged in order of increasing size?
(a)   solution, suspension, colloidal dispersion
(b)   solution, colloidal dispersion, suspension
(c)   colloidal dispersion, solution, suspension
(d)   colloidal dispersion, suspension, solution

32.   The degree of ionization of a compound can be determined by its:

(a)	temperature	(c)	electrical conductivity
(b)	color	(d)	odor

1. F **2.** T **3.** F **4.** F **5.** F **6.** T **7.** T **8.** F **9.** F **10.** T **11.** T **12.** T **13.** F **14.** d **15.** b **16.** c
17. c **18.** b **19.** (a) E (b) E (c) E (d) N **20.** (a) cation (b) anion (c) anion (d) cation
21. (a) Yes

$$(b) \quad S^{2-}_{(aq)} + 2\,Ag^+_{(aq)} \longrightarrow Ag_2S_{(s)}$$

22. (a) No
23. (a) Yes

$$(b) \quad Zn^{2+}_{(aq)} + CO_3^{2-}_{(aq)} \longrightarrow ZnCO_{3(s)}$$

24. (a) Yes

$$(b) \quad Ba^{2+}_{(aq)} + 2\,OH^-_{(aq)} \longrightarrow Ba(OH)_{2(aq)}$$

25. a **26.** c **27.** d **28.** b **29.** b **30.** a **31.** b **32.** c

Chapter 10   SOLUTION CONCENTRATIONS

The concentration of solute particles in a solution is often critical to the normal functioning of living organisms. In this chapter we study several methods used to measure the concentration of a solute in a solution.

10.1   Saturated and Unsaturated Solutions

The concentration of a solution indicates the number of solute particles dissolved in a certain amount of the solvent. If the solution is dilute, there are few solute particles; if it is concentrated, there are many. If it is saturated, the solvent contains all the solute particles it can hold at that temperature. Saturated solutions are not necessarily concentrated solutions. If a solute is very soluble in a particular solvent, the solution can be highly concentrated but still be unsaturated. If the solubility is very low, a very dilute solution may be saturated. A supersaturated solution is an unstable solution that results when the temperature of a saturated solution is lowered but no crystals form. Crystals will form very rapidly when a small crystal is added to a supersaturated solution.

Important Terms

> dilute   concentrated   saturated
> supersaturated   unsaturated

10.2   Molar Concentration

The concentration of a solution is a measure of the relative amount of solute in the solution, and is always expressed as a ratio. The molar concentration or molarity of a solution is defined as the number of moles of solute per liter of solution. The concentration of the majority of solutions you will use in the laboratory are expressed in molarity.

Important Term

> molarity

Example _____

1.   What does "0.5 M $MgCrO_4$" on a bottle label tell you about the solution in that bottle?

T   The solution will have 0.5 mole, or 70 grams, of $MgCrO_4$ dissolved in each liter of solution.

119

2.   How would you make 200 mL of 0.50 M KCl?

S   This problem asks how many grams of KCl must be weighed out and dissolved in
        enough water to make 200 mL of a 0.50 M solution.

T   0.50 M means 0.50 mole per liter of solution. To make one liter of solution you need
        0.50 mole of KCl. Determine the formula weight of KCl and multiply by 0.50.

E

$$0.50 \text{ mol KCl} \times \frac{74.6 \text{ g}}{1 \text{ mol KCl}} = 37.3 \text{ g KCl}$$

But the problem specifies only 200 mL of solution.

$$\frac{37.3 \text{ g}}{1000 \text{ mL}} \times 200 \text{ mL} = 7.5 \text{ g}$$

P   To make 200 mL of 0.50 M KCl you dissolve 7.5 grams of KCl in water and then add
        enough water to make 200 mL of solution.

3.   What is the molarity of 500 mL of solution that contains 1.7 grams of ammonia?

S   The question asks for the concentration in moles per liter of a solution that contains
        1.7 grams of NH_3 in 500 mL of solution.

T   The first step is to determine the number of moles of ammonia in 1.7 grams. Then
        divide the number of moles into the volume, in liters, to give you moles per liter (M).

E

$$1.7 \text{ g} \times \frac{1 \text{ mol } NH_3}{17 \text{ g}} = 0.10 \text{ mol of } NH_3$$

Our solution contains 0.10 mole in 500 mL.

$$M = \frac{\text{mol}}{\text{liter}} = \frac{0.10 \text{ mol}}{.500 \text{ L}} = 0.20 \text{ M}$$

P   The concentration is 0.20 M NH_3.

Check Your Understanding _____

For questions 1 to 4, calculate the molarity of the solution.
1.   9.0 grams of $C_6H_{12}O_6$ in 500 mL of solution.

2.   120 grams of NaOH in 2.0 liters of solution.
3.   0.14 mole of HCl/100 mL of solution.
4.   87.7 grams of NaCl in 750 mL of solution.

For questions 5 to 8, indicate how you would prepare each solution.
5.   0.800 liter of 2.45 M H_3PO_4
6.   375 mL of 0.250 M KOH
7.   100 mL of 6.00 M H_2SO_4
8.   0.25 liter of 0.10 M NaOH

9.   How many milliliters of 6.0 M HCl contain 0.40 mole of HCl?

10.3   Percent Concentration

Percent concentrations are units of concentration that do not take into account the formula weight of the solute. This means that you do not have to be concerned about the formula weight of the solute when calculating percent concentration. The problem is exactly the same no matter what solute is involved.

Two types of percent concentration are discussed in this section:

   Weight/volume (w/v) percent is defined as the number of grams of solute per 100 mL of solution.

   Milligram (mg%) percent is defined as the number of milligrams of solute per 100 mL of solution.

Important Terms

   weight/volume percent   milligram percent

Example   _____

How is 150 mL of 12.0 mg% NaCl solution prepared?

S   The problem is asking you to determine how many milligrams of NaCl must be used with enough water to make 150 ml of solution that is 12.0 Mg%.

T   There are two conversion factors that can be written.

$$\frac{12.0 \text{ mg NaCl}}{100 \text{ mL solution}} \quad \text{and} \quad \frac{100 \text{ mL solution}}{12.0 \text{ mg NaCl}}$$

121

E   Because we want an answer in mg, we use the first conversion factor

$$150 \text{ mL solution} \times \frac{12.0 \text{ mg NaCl}}{100 \text{ mL solution}} = 18.0 \text{ mg NaCl}$$

P   The calculated result is correct to three significant figures. To prepare the solution, 18.0 mg of NaCl is dissolved in enough water to make 150 mL of solution.

Check your Understanding _____

For questions 10 to 13, describe how you would prepare each solution.
10.   225 mL of 2.0% (w/v) glucose
11.   0.030 liter of 0.70% (w/v) $MgSO_4$
12.   5 deciliters of 8.0 mg% sucrose
13.   0.040 liter of 34 mg% $CaCl_2$

14.   What is the concentration in w/v percent of 500 mL of solution that contains 920 mg of $C_6H_{12}O_6$?
15.   What is the concentration of Na^+ in mg% if 5.0 mL of blood contains 0.14 mg of Na^+?
16.   What is the concentration of NaCl in w/v percent if 750 mL of solution contain 1.50 mole of NaCl?

For questions 17 to 25, fill in the blank with the correct word or words.
17.   A _____ solution contains very few solute particles, and a _____ solution contains many solute particles.
18.   In a _____ solution the solute particles in solution are in equilibrium with the undissolved solute particles in the container.
19.   A supersaturated solution can form when the temperature of a saturated solution is _____ and no crystals form.
20.   The unit of concentration that expresses concentration in moles per liter of solution is _____.
21.   _____ gives the concentration of a solution in grams of solute per 100 mL of solution.
22.   Concentration in _____ takes into account the formula weight of the solute.
23.   _____ concentrations do not take into account the formula weight of the solute.
24.   The concentrations of trace minerals in the blood are often expressed in units of _____.
25.   Clinical reports often have concentrations expressed in _____.

10.4   Parts per Million and Parts per Billion
Parts per million (ppm) and parts per billion (ppb) are units of concentration used to describe

extremely dilute solutions. You will find these terms used quite often to describe pollutants in water and air.

Important Terms

parts per million (ppm)   parts per billion (ppb)

Example ──

A sample of water was found to contain carbon tetrachloride in a concentration of 2 ppb. How many micrograms of carbon tetrachloride would be found in a 100 mL-sample?

S   This question asks: 100 mL water = (?) micrograms carbon tetrachloride.
T   There are two conversion factors given by the concentration.

$$\frac{2 \; \mu g}{1000 \; mL} \quad and \quad \frac{1000 \; mL}{2 \; \mu g}$$

Because the question asks for the mass of carbon tetrachloride in the sample we must use the first conversion factor.

E

$$\frac{2 \; \mu g}{1000 \; \cancel{mL}} \times 100 \; \cancel{mL} = 0.2 \; \mu g$$

P   A 100 mL sample contains 0.2 µg of carbon tetrachloride.

Check Your Understanding ────────────────────────────────

26.   A sample of waste water from a film processing plant contained silver in a concentration of 4 ppm. How many milligrams of silver would be found in a 200-mL sample?

27.   A 500-mL sample from a river in Japan was found to have 0.02 mg of cadmium. What is the concentration of cadmium in both parts per million and parts per billion?

10.5   Equivalents

The ionic components of a solution are often described in terms of equivalents. One equivalent is equal to 1 mole of charge. The equivalent weight of a substance is calculated by dividing the formula weight of the ion by the charge on the ion. The equivalent weight of an ion with a 1+ or 1- charge is the same as the formula weight. The equivalent weight of an ion with a 2+ or 2- charge is equal to one-half the formula weight. The ionic components of the blood are reported in milliequivalents (1000 mEq = 1 Eq).

Important Terms

equivalent (Eq)   equivalent weight   milliequivalent (mEq)

Example _____

A 100-mL sample of blood was found to contain 0.14 mEq of sodium ions. How many milligrams of sodium were there in the blood sample?

S   The question asks: 0.14 mEq Na^+ = (?) mg Na^+

T   To solve this problem you must determine the equivalent weight of sodium. You must then convert the equivalent weight expressed in grams per equivalent into milligrams per equivalent. Finally, you can calculate how many grams there are in 0.14 mEq.

E   The sodium ion is Na^+. 1 mol Na^+ contains 1 mol of + charge = 1 Eq

The equivalent weight of Na^+ = the weight of 1 mole = 23 g.

1 mEq of Na^+ will contain $\dfrac{23 \text{ g}}{1 \text{ Eq}} \times \dfrac{1 \text{ Eq}}{1000 \text{ mEq}} \times \dfrac{1000 \text{ mg}}{1 \text{ g}} = \dfrac{23 \text{ mg}}{1 \text{ mEq}}$

The blood sample contained $0.14 \text{ mEq} \times \dfrac{23 \text{ mg}}{1 \text{ mEq}} = 3.2 \text{ mg}$

P   0.14 mEq of sodium ions weigh 3.2 mg

Check Your Understanding _____

28.   How many milligrams of potassium ions are there in a sample of blood that contains 2.5 mEq of potassium ions?

29.   How many milliequivalents of magnesium ions are present in 75 mL of 0.80% (w/v) Mg^{2+} solution?

10.6 Dilutions

You will often be required to prepare a solution by diluting a more concentrated stock solution. The key to understanding dilutions to remember that only solvent is added. Although the concentration changes, the actual amount of solute remains the same when you dilute a solution. The equation:

$$C_2 \times V_2 = C_1 \times V_1$$

applies for all dilution problems as long as the same units of concentration and the same units of volume are used.

Example _____

How do you prepare 70 mL of 2% Na_2CO_3 from a 5% Na_2CO_3 stock solution?

S   The question asks how many mL of 5% Na_2CO_3 stock solution must be diluted with water to make 70 mL of 2% Na_2CO_3.

T   To solve this problem we substitute the given values into the dilution equation.

E $$2\% \; Na_2CO_3 \times 70 \; mL = 5\% \; Na_2CO_3 \times V$$

$$\frac{2\% \; Na_2CO_3 \times 70 \; mL}{5\% \; Na_2CO_3} = V$$

$$28 \; mL = V$$

P   To prepare the desired solution you measure out 28 mL of the stock solution, then add enough water to make 70 mL of solution.

Check Your Understanding _____

30.   How many milliliters of a 6.0 M HBr stock solution would you need to prepare 15 mL of 2.5 M HBr?

31.   How many milliliters of a 10% NaCl solution would you need to prepare 0.50 liters of physiological saline (0.90% NaCl)?

32.   What is the final concentration of a solution made from a 1:4 dilution of 3.6 M NaOH?

10.7   Colligative Properties

Properties of solutions that depend only upon the concentration of solute particles in the solution are called colligative properties. The colligative properties include the raising of the boiling point and the lowering of the freezing point of water by solute particles.

Important Term

colligative property

10.8 Osmosis

Another colligative property is osmotic pressure. Osmosis is the migration of water molecules through a differentially permeable membrane (one that will let water, but not solute particles, pass through) from a region of lower solute concentration to a region of higher solute concentration. Osmotic pressure is defined as the amount of pressure that must be applied to prevent this flow of water. The higher the solute concentration, the greater the osmotic pressure. Osmosis can be seen as an attempt to achieve equal osmotic pressures on both sides of the membrane.

The osmolarity of a solution indicates the total number of moles of all solute particles in one liter of solution. One osmol is one mole of any combination of solute particles (molecules or ions). Solutions having different types of solute particles but the same osmolarity will have the same osmotic pressure. The higher the osmolarity of a solution, the greater the osmotic pressure of that solution.

Important Terms

      osmosis   osmotic pressure   osmolarity
      osmol (osm)   milliosmol (mOsm)

Example   ———————————————————————————————

What is the osmolarity of a one liter solution containing one mole of calcium nitrate and one mole of glucose?

S   The question asks you to determine how many moles of particles (osmoles) there are in a one liter solution that contains one mole of calcium nitrate and one mole of glucose.

T•E   1 mole of $Ca(NO_3)_2$ will dissolve in solution to form 1 mole of Ca^{2+} ions and 2 moles of NO_3^- ions. The glucose molecule doesn't ionize when dissolved in water, so there will be 1 mole of glucose molecules in solution. The total number of moles of particles in solution is four.

P   The osmolarity of the solution is 4 osmols/liter.

10.9 Isotonic Solutions

The movement of water is critical to living organisms. If the concentration of the solution outside a cell is equal to the concentration inside the cell, the solution is said to be isotonic. In an isotonic solution, water will move in and out of the cell at the same rate. If the concentration of the solution outside the cell is greater than that inside the cell, the solution is hypertonic to the cell; water will flow out of the cell, causing the cell to shrink. If the concentration of the solution outside the cell is less than that inside the cell, the solution is hypotonic to the cell; water will move into the cell, causing the cell to swell. Electrolytes in body fluids are important in maintaining the osmotic balance in the body.

Important Terms

| isotonic | hypertonic | hypotonic | edema |
| hemolysis | crenation | normal saline | |

Check Your Understanding ——————————————————————

For questions 33 to 39, answer true (T) or false (F).

33. Osmosis is the movement of water through a membrane from a region of low solute concentration to a region of higher solute concentration.
34. Solutions whose osmotic pressures are equal are said to be isotonic.
35. Red blood cells placed in a 0.2 M NaCl solution will undergo crenation.
36. Red blood cells placed in 200 mL of solution containing 40 grams of NaCl will undergo hemolysis.
37. No change will be observed in red blood cells placed in 0.5 liter of solution containing 4.5 grams of NaCl.
38. A solution of 2 M K_3PO_4 will have an osmolarity of 4 osmol/liter.
39. 100 mL of 3 M NaCl will have the same osmotic pressure as 100 mL of 3 M NaOH.

For questions 40 to 49 select the best answer.

40. A solution that contains all of the solute that it can under the given conditions is:
  (a) concentrated   (c) dilute
  (b) saturated   (d) unsaturated

41. The flow of water molecules through a differentially permeable membrane from a region of lower solute concentration to a region of higher solute concentration is called:
  (a) osmolarity   (c) hypertonic
  (b) osmosis   (d) hypotonic

42. If the concentration of solute is equal on both sides of a membrane, the solutions are said to be:
  (a) hypotonic   (c) isotonic
  (b) hypertonic   (d) normal

43. A red blood cell will undergo which of these changes when immersed in a hypertonic solution?
  (a) crenation   (c) hemolysis
  (b) swelling   (d) no effect

127

44. A 5.00 L solution that contains 1.75 mole of solute is:
   - (a) 1.75 M
   - (b) 8.75 M
   - (c) 5.00 M
   - (d) 0.35 M
45. What mass of NaBr is needed to prepare 5.00 L of 0.020 M solution?
   - (a) 1030 g
   - (b) 10.3 g
   - (c) 20.6 g
   - (d) 2.06 g
46. What is the osmolarity of a 0.50 M aqueous Na_2SO_4 solution?
   - (a) 0.05 osm/L
   - (b) 1.50 osm/L
   - (c) 1.00 osm/L
   - (d) 2.00 osm/L
47. Which of the following aqueous solutions has the highest osmotic pressure?
   - (a) 1.0 M LiCl
   - (b) 0.5 M sugar
   - (c) 1.0 M K_2CO_3
   - (d) 0.5 M Na_3PO_4
48. If 7.00×10^2 mL of aqueous solution contains 50.0 g of NaCl, the concentration of the solution is:
   - (a) 2.62% (w/v)
   - (b) 3.33% (w/v)
   - (c) 1.31% (w/v)
   - (d) 7.14% (w/v)
49. If 25.0 mL of a 0.08 M HCl solution is diluted to 50.0 mL, what is the new concentration of the solution?
   - (a) 0.04 M
   - (b) 2.0 M
   - (c) 0.16 M
   - (d) 4.0 M

Answers to Check Your Understanding Questions in Chapter 10

1. $9.0 \cancel{g} \times \dfrac{1\ mol}{180\ \cancel{g}} = 0.050\ mol$

$$\dfrac{0.050\ mol}{500\ \cancel{mL}} \times \dfrac{1000\ \cancel{mL}}{1\text{-liter}} = \dfrac{0.10\ mol}{1\text{-liter}} = 0.10\ \text{M glucose}$$

2. $120 \cancel{g} \times \dfrac{1\ mol}{40\ \cancel{g}} = 3.0\ mol;\quad \dfrac{3.0\ mol}{2.0\ liter} = \dfrac{1.5\ mol}{1\ liter} = 1.5\ \text{M NaOH}$

3. $\dfrac{0.14\ mol}{100\ \cancel{mL}} \times \dfrac{1000\ \cancel{mL}}{1\ liter} = \dfrac{1.4\ mol}{1\ liter} = 1.4\ \text{M HCl}$

4. $87.7 \cancel{g} \times \dfrac{1\ mol}{58.5\ \cancel{g}} = 1.50\ mol$

$$\frac{1.50 \text{ mol}}{750 \text{ mL}} \times \frac{1000 \text{ mL}}{1 \text{ liter}} = \frac{2.00 \text{ mol}}{1 \text{ liter}} = 2.00 \text{ M NaCl}$$

5. $\quad 0.800 \text{ liter} \times \dfrac{2.45 \text{ mol } H_3PO_4}{1 \text{ liter}} \times \dfrac{98.0 \text{ g}}{1 \text{ mol } H_3PO_4} = 192 \text{ g}$

Add 192 g of H_3PO_4 to enough water to make 0.800 liter of solution.

6. $\quad 375 \text{ mL} \times \dfrac{0.250 \text{ mol KOH}}{1000 \text{ mL}} \times \dfrac{56.1 \text{ g}}{1 \text{ mol KOH}} = 5.26 \text{ g}$

Add 5.26 g of KOH to enough water to make 375 mL of solution.

7. $\quad 100 \text{ mL} \times \dfrac{6.00 \text{ mol } H_2SO_4}{1000 \text{ mL}} \times \dfrac{98.1 \text{ g}}{1 \text{ mol } H_2SO_4} = 58.9 \text{ g}$

Add 58.9 g of H_2SO_4 to enough water to make 100 mL of solution.

8. $\quad 0.25 \text{ liter} \times \dfrac{0.10 \text{ mol NaOH}}{1 \text{ liter}} \times \dfrac{40 \text{ g}}{1 \text{ mol NaOH}} = 1.0 \text{ g}$

Add 1.0 g NaOH to enough water to make 0.25 liter of solution.

9. $\quad 0.40 \text{ mol} \times \dfrac{1 \text{ liter}}{6.0 \text{ mol}} \times \dfrac{1000 \text{ mL}}{1 \text{ liter}} = 67 \text{ ml}$

10. $\quad 225 \text{ mL} \times \dfrac{2.0 \text{ g glucose}}{1000 \text{ mL}} = 4.5 \text{ g glucose}$

Add 4.5 g glucose to enough water to make 225 mL solution.

11. $\quad 0.030 \text{ liter} \times \dfrac{1000 \text{ mL}}{1 \text{ liter}} \times \dfrac{0.70 \text{ g } MgSO_4}{100 \text{ mL}} = 0.21 \text{ g } MgSO_4$

Add 0.21 g $MgSO_4$ to enough water to make 0.030 liter of solution.

12. $\quad 5 \text{ dL} \times \dfrac{100 \text{ mL}}{1 \text{ dL}} \times \dfrac{8 \text{ mg sucrose}}{100 \text{ mL}} = 40 \text{ mg sucrose}$

Add 40 mg of sucrose to enough water to make 5 dL of solution.

13. $0.040 \text{ liter} \times \dfrac{1000 \text{ mL}}{1 \text{ liter}} \times \dfrac{34 \text{ mg CaCl}_2}{100 \text{ mL}} = 14 \text{ mg CaCl}_2$

Add 14 mg $CaCl_2$ to enough water to make 0.040 liter of solution.

14. $920 \text{ mg} \times \dfrac{1 \text{ g}}{1000 \text{ mg}} \; ; \quad \dfrac{0.920 \text{ g C}_6\text{H}_{12}\text{O}_6}{500 \text{ mL}} = \dfrac{X}{100 \text{ mL}}$

$$X = 0.184 \text{ g}$$

$$\dfrac{0.184 \text{ g}}{100 \text{ mL}} = 0.184\% \text{ glucose}$$

15. $\dfrac{0.14 \text{ mg Na}^+}{5.0 \text{ mL}} = \dfrac{X}{100 \text{ mL}} \; ; X = 2.8 \text{ mg} \; ; 2.8 \text{ mg\%}$

16. $1.50 \text{ mol} \times \dfrac{58.5 \text{ g NaCl}}{1 \text{ mol}} = 87.8 \text{ g}; \quad \dfrac{87.8 \text{ g NaCl}}{750 \text{ mL}} = \dfrac{X}{100 \text{ mL}}$

$$X = 11.7\% \text{ NaCl}$$

17. dilute, concentrated 18. saturated 19. lowered 20. molarity 21. weight/volume percent
22. molarity 23. percent 24. milligram percent 25. weight/volume or milligram percent

26. $200 \text{ mL} \times \dfrac{4 \text{ mg Ag}}{1 \text{ liter}} \times \dfrac{1 \text{ liter}}{1000 \text{ mL}} = 0.8 \text{ mg Ag}$

27. $\dfrac{0.02 \text{ mg Cd}}{500 \text{ mL}} \times \dfrac{1000 \text{ mL}}{1 \text{ liter}} = \dfrac{0.04 \text{ mg Cd}}{1 \text{ liter}} = 0.04 \text{ ppm Cd}$

$\dfrac{0.04 \text{ mg Cd}}{1 \text{ liter}} \times \dfrac{1000 \text{ μg}}{1 \text{ mg}} = \dfrac{40 \text{ μg Cd}}{1 \text{ liter}} = 40 \text{ ppb Cd}$

28. $2.5 \text{ mEq K}^+ \times \dfrac{39.1 \text{ g}}{1 \text{ Eq}} \times \dfrac{1 \text{ Eq}}{1000 \text{ mEq}} \times \dfrac{1000 \text{ mg}}{1 \text{ g}} = 98 \text{ mg K}^+$

29. $0.80\% \ Mg^{2+} = \dfrac{0.80 \ g \ Mg^{2+}}{100 \ mL} \times 75 \ mL = 0.60 \ g \ Mg^{2+}$

$0.60 \ g \ Mg^{2+} \times \dfrac{1 \ Eq}{12.2 \ g} \times \dfrac{1000 \ mEq}{1 \ Eq} = 49 \ mEq \ Mg^{2+}$

30. $2.5 \ M \times 15 \ mL = 6.0 \ M \times V; \quad V = 6.3 \ mL$

31. $0.50 \ liter \times 0.90\% \ NaCl = 10\% \ NaCl \times V$

$V = 0.045 \ liter = 45 \ mL$

32. $4V \times C = 1V \times 3.6 \ M \ NaOH; \quad C = 0.90 \ M \ NaOH$

33. T **34.** T **35.** T **36.** F **37.** T **38.** F **39.** T

40. b **41.** b **42.** c **43.** a **44.** d **45.** b **46.** b **47.** c **48.** d **49.** a

Chapter 11   ACIDS AND BASES

In this chapter we study two very important classes of compounds: acids and bases. We discuss their properties, their interactions, the methods used to describe and measure their concentrations, and the way in which living organisms protect themselves against large changes in the concentrations of acid and base.

11.1   Acids and Bases: the Brønsted-Lowry Definition

Acids and bases play an important role in body chemistry. Acids form aqueous solutions that taste sour, turn litmus paper red, and react with metals to produce hydrogen. Bases form aqueous solutions that taste bitter, feel slippery to the touch, and turn litmus paper blue.

There are several definitions of acids and bases, but we will use the Brønsted-Lowry definition. An acid is a substance that donates protons (the positive hydrogen ion, H^+), and a base is a substance that accepts protons. In this definition we also talk about conjugate acid-base pairs. The conjugate base of an acid is the negative ion that results when the acid donates a hydrogen ion. The difference between an acid and its conjugate base is always one proton. Consider the following equation:

$$H_2CO_3 \; + \; H_2O \; \rightleftharpoons \; H_3O^+ \; + \; HCO_3^-$$
$$\text{acid}_1 \qquad \text{base}_2 \qquad\qquad \text{acid}_2 \qquad \text{base}_1$$

The bicarbonate ion (HCO_3^-) is the conjugate base of carbonic acid (H_2CO_3.) The hydronium ion is the conjugate acid of water, which is acting as a base in this reaction.

Binary acids (acids containing hydrogen and one other nonmetallic element) are named by adding the prefix, hydro-, to the name of the second element. The name is given an -ic ending and then the word "acid" is added. The names of salts formed from binary acids end in -ide. For example, HCl is hydrochloric acid and its sodium salt is NaCl, sodium chloride.

Important Terms

acid	base	conjugate acid-base pair
hydronium ion	hydroxide ion	polyprotic acid
amphoteric	binary acid	

Check Your Understanding _____

1.   List three characteristics of (a) acidic solutions and (b) basic solutions.

132

For questions 2 to 5, identify the two conjugate acid-base pairs in each:

2. $HF + NH_3 \rightleftharpoons NH_4^+ + F^-$

3. $N_2H_4 + H_2O \rightleftharpoons N_2H_5^+ + OH^-$

4. $H_2SO_4 + H_2O \rightleftharpoons H_3O^+ + HSO_4^-$

5. $H_2SO_3 + HNO_3 \rightleftharpoons H_3SO_3^+ + NO_3^-$

11.2 Strength of Acids and Bases

A strong acid is one that readily donates a proton. If an acid readily donates a proton, its conjugate base will have a weak attraction for protons. Strong acids have weak conjugate bases. A weak acid does not readily donate a proton. If an acid does not readily donate a proton, its conjugate base has a strong attraction for protons. Weak acids have strong conjugate bases. The same is true for bases. A strong base has a large attraction for protons and its conjugate acid will be weak. A weak base will not attract protons as readily, and its conjugate acid will be strong.

Check Your Understanding

6. Identify each of the following as an acid or a base, then list the acids and the bases in order of increasing strength:

OH^-, NO_3^-, H_2O, H_2SO_4, HPO_4^{2-}, PO_4^{3-}, HCl, HNO_3, H_2CO_3, CO_3^{2-}

11.3 Neutralization Reactions

An acid will react with a base to produce water and a salt. If a solution contains equal amounts of hydrogen ions and hydroxide ions it is said to be neutral. As a result, it will have neither acidic nor basic properties. A neutralization reaction is one in which a solution of acid reacts with a solution of base to produce a neutral solution.

Important Terms

  neutral solution   neutralization   salt

Check Your Understanding

7. Write the balanced equation and the net-ionic equation for:
   (a)   The neutralization of $Ca(OH)_2$ by HCl.
   (b)   The neutralization reaction between HNO_3 and KOH.

11.4   Ionization of Water

Water molecules can be ionized by other water molecules to form the hydrogen ion and the hydroxide ion. Although we use the term hydrogen ion or proton when talking about acids and bases, the hydrogen ion does not exist as a separate particle in aqueous solutions. It is always joined to a water molecule in the form of the hydronium ion, H_3O^+. The hydronium ion contains a coordinate covalent bond in which the oxygen donates both electrons found in the bond between the hydrogen and the oxygen.

$$H_2O + H_2O \rightleftharpoons \underset{\text{hydronium ion}}{H_3O^+} + \underset{\text{hydroxide ion}}{OH^-}$$

Important Terms

hydrogen ion   hydroxide ion
coordinate covalent bond   hydronium ion

11.5   Ion Product of Water, K_w

In pure water the concentration of the hydrogen ion times the concentration of the hydroxide ion equals a constant called the ion product of water, K_w. The numerical value of K_w is 1×10^{-14}. Therefore,

$$[H^+][OH^-] = 1 \times 10^{-14}$$

If you know either $[H^+]$ or $[OH^-]$ in a solution you can calculate $[OH^-]$ or $[H^+]$ using the expression for K_w.

Important Term

ion product of water, K_w

Example _____

If the concentration of hydroxide ions in a solution is 0.00082 M, what is the hydrogen ion concentration?

S   The question asks: $[H^+]$ = (?) Mol/L.

T   From the statement of the problem we have $[OH^-]$. We can use the equation for the ion product of water to determine $[H^+]$.

E $$[H^+] = \frac{1 \times 10^{-14}}{[OH^-]} = \frac{1 \times 10^{-14}}{0.0082} = 1.2 \times 10^{-12}\,M$$

P   The answer should have two significant figures and is 1.2×10^{-12} M

Check Your Understanding _____

Calculate the hydroxide ion concentration in moles per liter for solutions whose hydrogen ion concentration is:
8. 1×10^{-8} M
9.   0.0066 M
10. 5.5×10^{-5} M
11.   0.0065 mol in 600 mL

12.7   The pH Scale

The molarities of solutions of acids or bases found in living organisms normally fall within certain ranges. A scale called the pH scale was devised as an easy way to show the concentration of the hydrogen ion in solution. pH is defined as minus the log of the hydrogen ion concentration.

$$pH = - \log [H^+]$$

Pure water, which is neutral, has a pH of 7. Study carefully Table 11.3 for the conversions between concentration and pH. Figure 11.1 lists the pH of some important common substances and Table 11.4 gives the pH range for several body fluids.

The concentration of an acid or base solution can be determined using a procedure called a titration. In a titration, an acidic or basic solution of known concentration is added to a solution of unknown concentration until equal amounts of acid and base are in the container. That point, called the equivalence point, is determined by an acid-base indicator or a pH meter. From the number of milliliters of known solution added, and from the known concentration of this acid or base, the concentration of the unknown base or acid can be determined. Table 11.5 lists several acid-base indicators.

Important Terms

pH	pH meter	acid-base indicator
titration	end point	equivalence point

Check Your Understanding _____

For questions 12 to 18, give the (a) $[H^+]$, (b) $[OH^-]$, (c) pH, and (d) indicate if the solution is acidic or basic.
12.   0.1 M HCl

13. 1×10^{-4} M HNO_3
14. 0.001 M NaOH
15. 1×10^{-6} M KOH
16 0.0001 M lithium hydroxide
17. 36.5 µg HCl in 100 mL of solution
18. 370 mg of $Ca(OH)_2$ in one liter of solution. (Hint: Each mole of $Ca(OH)_2$ yields 2 moles of OH⁻ in solution.)

11.7 What Are Buffers?

Buffers are substances that, when placed in a solution, protect it against changes in pH. The best buffer systems consist of a weak acid and its conjugate base, or a weak base and its conjugate acid. These systems have their greatest buffering capacity at that pH where the concentration of the acid equals the concentration of its conjugate base.

Important Terms

buffer   acidosis   alkalosis

Check Your Understanding _____

For questions 19 to 22, choose the best answer or answers from the choices given.
19. Substances that protect the blood from large changes in pH are called
    (a)   electrolytes   (c)   acids
    (b)   bases   (d)   buffers
20. A drop in the alkaline reserves of the blood will cause a condition called
    (a)   acidosis   (c)   alkalosis
    (b)   edema   (d)   emphysema
21. Hyperventilation can result in a change in blood pH, causing a condition called
    (a)   acidosis   (c)   alkalosis
    (b)   diabetes mellitus   (d)   congestive heart failure
22. The best buffer systems are composed of
    (a)   a strong acid and its conjugate base
    (b)   a weak acid and its conjugate base
    (c)   a strong base and its conjugate acid
    (d)   a weak base and its conjugate acid

23. Describe the effects of (a) adding acid and (b) adding base to an aqueous solution containing the following buffer system:

$$NH_4OH + H^+ \rightleftharpoons NH_4^+ + H_2O$$

136

For questions 24 to 31 select the best answer.

24. A buffer maintains
     (a) a pH of 7 all of the time
     (b) a pH greater than 7 all of the time
     (c) a pH less than 7 all of the time
     (d) a constant pH when small amounts of acid or base are added

25. Which of the following is not a characteristic of bases?
     (a) turn litmus blue   (c) taste sour
     (b) feel slippery   (d) neutralize acids

26. Litmus changes from blue to red in beer. Therefore beer is
     (a) neutral   (c) acidic
     (b) basic   (d) sudsy

27. Which of the following is not acidic?
     (a) sauerkraut   (c) household ammonia
     (b) gastric juice   (d) lemon juice

28. Most amines ionize only very slightly in water, producing relatively few hydroxide
     ions. Amines are:
     (a) weak acids   (c) strong bases
     (b) weak bases   (d) strong acids

29. Which of the following is the conjugate base of $H_2PO_4^{2-}$?
     (a) H_3PO_4 (c) HPO_4^-
     (b) PO_4^{3-} (d) OH^-

30. If $[H^+]$ is 1×10^{-3}, the pH of the solution is:
     (a) 1   (c) 3
     (c) 7   (d) 11

31. If the pH of a solution is 4, then $[OH^-]$ is:
     (a) $1 \times 10^{-4}\,M$ (c) $1 \times 10^{+4}\,M$
     (b) $1 \times 10^{-10}\,M$ (d) $1 \times 10^{+10}\,M$

Answers to Check Your Understanding Questions in Chapter 11

1. (a) React with metals to produce hydrogen, taste sour, turn litmus paper red, neutralize bases (b) Taste bitter, feel slippery, turn litmus paper blue, neutralize acids **2.** HF/F^- and NH_4^+/NH_3 **3.** H_2O/OH^- and $N_2H_5^+/N_2H_4$ **4.** H_2SO_4/HSO_4^- and H_3O^+/H_2O **5.** HNO_3/NO_3^- and $H_3SO_3^+/H_2SO_3$ **6.** (a) Acids in order of increasing strength: H_2O, HPO_4^{2-}, H_2CO_3, HNO_3, HCl, H_2SO_4 (b) Bases in order of increasing strength: NO_3^-, H_2O, HPO_4^{2-}, CO_3^{2-}, PO_4^{3-}, OH^-

7.

(a) $Ca(OH)_2 + 2HCl \longrightarrow 2H_2O + CaCl_2$

   $2OH^- + 2H^+ \longrightarrow 2H_2O$

(b) $HNO_3 + 2H^+ \longrightarrow H_2O + KNO_3$

   $H^+ + OH^- \longrightarrow H_2O$

8. $[OH^-] = \dfrac{1 \times 10^{-14}}{1 \times 10^{-8}} = 1 \times 10^{-6}\,M$

9. $[OH^-] = \dfrac{1 \times 10^{-14}}{.0066} = 1.5 \times 10^{-12}\,M$

10. $[OH^-] = \dfrac{1 \times 10^{-14}}{5.5 \times 10^{-5}} = 1.8 \times 10^{-10}\,M$

11. $[H^+] = \dfrac{.0065\ mol}{.600\ L} = 1.1 \times 10^{-2}\,M$

   $[OH^-] = \dfrac{1 \times 10^{-14}}{1.1 \times 10^{-2}} = 9.1 \times 10^{-13}\,M$

12. (a) $[H^+] = 0.1\,M$
   (b) $[OH^-] = \dfrac{1 \times 10^{-14}}{[H^+]} = \dfrac{1 \times 10^{-14}}{1 \times 10^{-1}} = 1 \times 10^{-13}\,M$

   (c) $pH = 1$
   (d) acidic

13. (a) $[H^+] = 1 \times 10^{-4}\,M$
   (b) $[OH^-] = 1 \times 10^{-10}\,M$

   (c) $pH = 4$
   (d) acidic

14. (a)
   $[H^+] = \dfrac{1 \times 10^{-14}}{[OH^-]} = \dfrac{1 \times 10^{-14}}{1 \times 10^{-3}} = 1 \times 10^{-11}\,M$
   (b) $[OH^-] = 0.001\,M$

   (c) $pH = 11$
   (d) basic

15. (a) $[H^+] = 1 \times 10^{-8}$
   (b) $[OH^-] = 1 \times 10^{-6}$

   (c) $pH = 8$
   (d) basic

16. (a) $[H^+] = 1 \times 10^{-10}$
   (b) $[OH^-] = 1 \times 10^{-4}$

   (c) $pH = 10$
   (d) basic

17. (a)

$$[H^+] = \frac{36.5\ \mu g}{100\ mL} \times \frac{1\ g}{1 \times 10^6\ \mu g} \times \frac{1\ mol}{36.5\ g} \times \frac{1000\ mL}{1\ liter} = \frac{1 \times 10^{-5}\ mol}{1\ liter}$$

(b) $[OH^-] = 1 \times 10^{-9}$ (c) pH = 5 (d) acidic

18. (a) $[H^+] = 1 \times 10^{-12}$ M

(b) $$\frac{370\ mg\ Ca(OH)_2}{1\ liter} \times \frac{1\ g}{1000\ mg} \times \frac{1\ mol}{74.1\ g} = \frac{0.005\ mol\ Ca(OH)_2}{1\ liter}$$

$[OH^-] = 0.01$ M $= 1 \times 10^{-2}$ M

(c) pH = 12 (d) basic

19. d **20.** a **21.** c **22.** b, d **23.** (a) drive the reaction to the right, producing more NH_4^+ and H_2O (b) drive the reaction to the left, producing more NH_4OH and H^+
24. d **25.** c **26.** c **27.** c **28.** b **29.** c **30.** c **31.** b

Chapter 12 CARBON AND HYDROGEN: THE HYDROCARBONS

This chapter begins our study of the chemistry of carbon compounds. It introduces new vocabulary, new methods for naming compounds, and new ways of writing chemical formulas that describe the structure of these compounds. All this may seem a bit complex at first; however, as was true of Chapter 1, a little extra time spent mastering these new concepts will make it much easier to read and study the rest of the book.

12.1 Organic Chemistry: An Overview

Carbon is unique among the elements in forming strong stable bonds with up to four other carbon atoms. It does so because of its position in Group IVA on the periodic table. Molecules containing carbon atoms can form long chains, branched chains, and rings. These carbon compounds can form molecules (called isomers) having identical molecular formulas but different geometric structures.

Besides forming stable bonds with other carbon atoms, carbon can form stable bonds with many other elements such as hydrogen, oxygen, nitrogen, sulfur, phosphorus, and the halogens. This results in an enormous number of different compounds and has led to a separate field of chemistry, called organic chemistry, devoted to the study of carbon compounds.

In order to study organic chemistry you will need to learn to recognize functional groups. These are particular arrangements of atoms that give rise to a certain set of reactions. For each functional group, you will learn several reactions that are of particular importance for living organisms. These reactions will show up again in Section III of the text when you learn about biochemical processes in the body.

As you learn each functional group, you will also learn how to name compounds that contain the group. Because there are so many organic compounds, there must be a system for naming them. The system in use throughout the world is that developed by the International Union of Pure and Applied Chemistry, called the IUPAC system.

Important Terms

      organic chemistry   functional group   IUPAC

12.2 Bonding with Hybrid Orbitals

Covalent bonds are formed by the overlap of atomic orbitals: the greater the overlap, the stronger the bond. In forming some bonds, atomic orbitals will mix or hybridize to form hybrid orbitals, whose shapes allow for greater overlap. In order to form four equal covalent bonds, the single 2s and the three 2p orbitals in carbon hybridize and form four equal sp^3 orbitals, each directed toward the corner of a tetrahedron. Because of this hybridization, when the carbon atom forms four single bonds, the resulting molecule is tetrahedral in shape.

Important Terms

hybridization   hybrid orbital sp^3 orbital

12.3 Structural Formulas and Isomers

Structural formulas are chemical diagrams that show the position of the atoms in a molecule. Condensed structural formulas show the position of the atoms without drawing all the bond lines. Practice drawing extended structural formulas from condensed formulas. Work your way from one end of the molecule to the other keeping in mind that every carbon must have four bonds. Also practice writing condensed formulas when you are given an extended structural formula.

C_5H_{12} is a molecular formula. It tells us how many carbons and hydrogens there are in the compound, but it does not tell us how they are arranged.

This is a structural formula that corresponds to the molecular formula C_5H_{12}. It is also known as an extended structural formula.

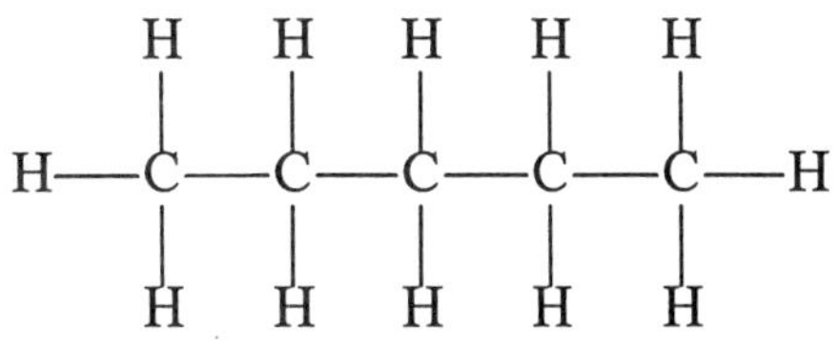

Sometimes condensed structural formulas are used. This is the condensed structural formula that corresponds to the extended structural formula.

$$CH_3CH_2CH_2CH_2CH_3$$

Both of these methods of writing the structural formula give us the same information about the compound. They tell us the bonding arrangement in the compound.

There are two other compounds that have the same molecular formula (C_5H_{12}) but different structural formulas than shown above. The extended and condensed structural formulas for these two compounds are:

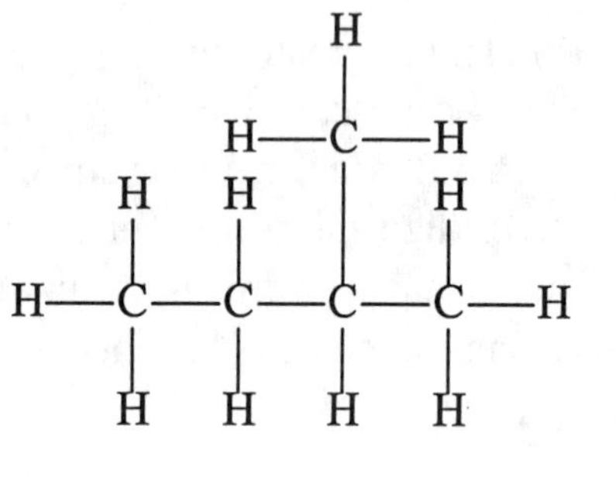

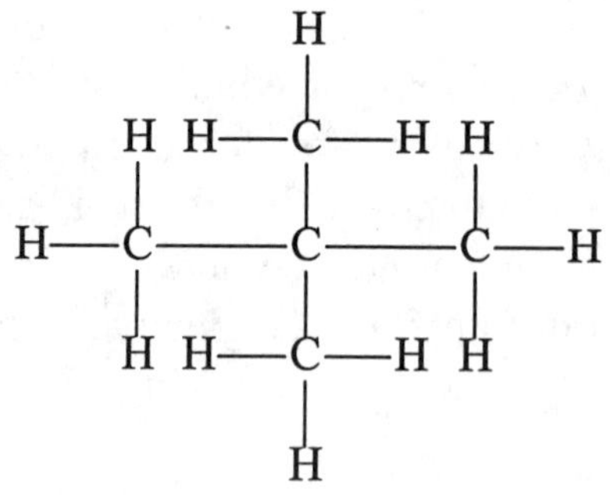

$$CH_3CH_2CH(CH_3)_2$$

$$CH_3C(CH_3)_2CH_3$$

Important Terms

structural formula   condensed structural formula

Example _____

There are two isomers that have the molecular formula C_4H_{10}. We can write their structural formulas as follows:

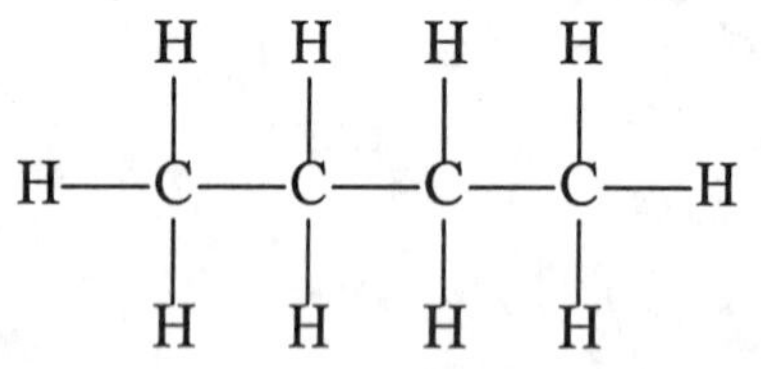

Isomer 1 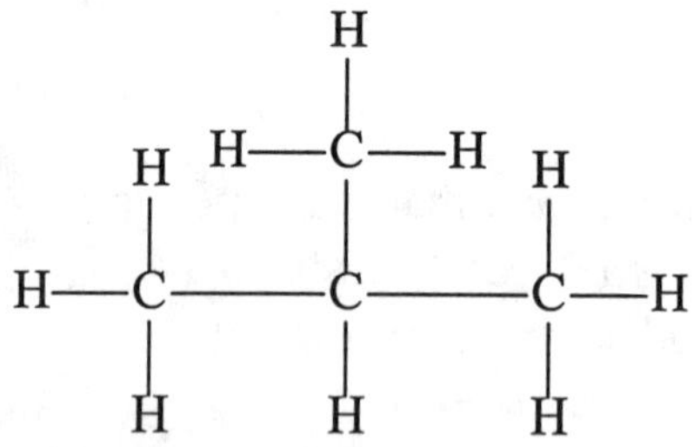   Isomer 2

1.   Write the condensed structural formula for isomer 1.

S   The extended structural formula must be converted into the condensed structural formula.

T•E   Start from one end of the compound and write each carbon and the atoms attached to it until you have done so for all of the carbons.

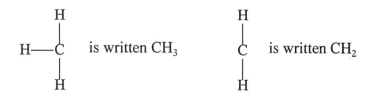

P   The condensed structural formula is $CH_3CH_2CH_2CH_3$, which can also be written as $CH_3(CH_2)_2CH_3$.

2.   Write the condensed structural formula for isomer 2.

S   This problem is the same, write the condensed structural formula.

T•E   Use the same procedure as you used to answer the previous question. Groups that are attached to the carbon chain are shown in parentheses.

$$-\overset{|}{\underset{|}{C}}-H \qquad \text{is written CH}$$

P   The condensed structural formula for this compound is

$$CH_3CH(CH_3)CH_3$$

Check Your Understanding _____

For questions 1 to 4, write the condensed structural formula for the given structural formula.

1.
$$H-\overset{\overset{\displaystyle H}{|}}{\underset{\underset{\displaystyle H}{|}}{C}}-\overset{\overset{\displaystyle H}{|}}{\underset{\underset{\displaystyle H}{|}}{C}}-\overset{\overset{\displaystyle H}{|}}{\underset{\underset{\displaystyle H}{|}}{C}}-\overset{\overset{\displaystyle H}{|}}{\underset{\underset{\displaystyle H}{|}}{C}}-\overset{\overset{\displaystyle H}{|}}{\underset{\underset{\displaystyle H}{|}}{C}}-\overset{\overset{\displaystyle H}{|}}{\underset{\underset{\displaystyle H}{|}}{C}}-H$$

2.

3.

```
              H
              |
        H—C—H
   H          H       H
   |          |       |
H—C————C——C————C—H
   |          |       |
   H          H       H
        H—C—H
              |
              H
```

4.

```
                      H
                      |
              H—C—H
              H       H   H
              |       |   |
        H—C——C——C==C—H
              |   |
              H   H
```

For questions 5 to 8, write the complete structural formula for the given condensed structural formula.

5. $CH_3CH{=}CHC(CH_3)_2CH_3$
6. $CH_3CH_2CHOHCH_2CH_3$
7. $CH_3CH_2CH(CH_3)CH_2Cl$
8. $CH_3CH(C_2H_5)CH_2CH(CH_3)CH_3$
9. Write the structural formulas for the nine isomers with molecular formula C_7H_{16}.

12.4 Alkanes: The Saturated Hydrocarbons

Hydrocarbons are a large class of organic compounds whose molecules contain only hydrogen and carbon. Petroleum is the major source of hydrocarbon compounds. The different hydrocarbons in petroleum are separated by a process called fractional distillation. This process makes use of the fact that compounds with longer hydrocarbon chains have higher boiling points than compounds with shorter chains.

The alkanes are a class of hydrocarbons whose molecular formulas fit the general formula C_nH_{2n+2}. The chief characteristic of the alkanes is that all the bonds in the compound are single bonds. Compounds that contain all carbon-to-carbon single bonds are said to be saturated. As a result, alkanes are the least reactive class of hydrocarbons.

Important Terms

alkane	saturated hydrocarbon	fractional distillation
hydrocarbon	petroleum	

12.5 Constitutional Isomers

Molecular formulas tell us what atoms are present in a molecule and how many of each atom are present. Constitutional isomers are compounds having identical molecular formulas, but different three-dimensional arrangements of atoms in their molecules. Although these isomers have the same number of atoms of each element, the atoms are attached differently. This gives the isomers different chemical and physical properties. The number of possible constitutional isomers increases as the number of carbon atoms in the compound increases.

Be systematic as you try to decide the relationship between structural formulas. First determine if the two structures have the same atoms and the same number of these atoms. If they do not, they are not isomers. They are unrelated compounds. If the two structures do have the same atoms and the same number of these atoms, determine if the atoms are connected in the same way. Be especially careful not to be confused by identical compounds that are flipped horizontally or vertically. Review problems 13 and 14 in the text provide numerous examples for your use in improving your skill at identifying the relationship between structures.

Important Term

constitutional isomer

12.6   Nomenclature

The IUPAC rules give a standardized procedure for naming organic compounds. Study carefully the basic rules listed in Section 12.6 of your text.

Example _____

1.   Name the following compound:

$$CH_2CH_3$$
$$|$$
$$CH_3CH_2CHCH_2CH_2CHCH_3$$
$$|$$
$$CH_3$$

S   The IUPAC rules for naming compounds must be applied to this compound.

T•E   (a)   All the bonds in this compound are single bonds so it is an alkane and the name will end in -ane.

(b)   Count the number of carbons in the longest carbon chain. There are seven. The prefix that means seven is hept-. This compound is a heptane.

(c)   A methyl group ($-CH_3$) and an ethyl group ($-CH_2CH_3$) are attached to the chain.

145

(d)   The name of each attached group must be preceded by a number that indicates its position on the carbon chain. In this case, there are two possibilities: 3-ethyl-6-methyl or 5-ethyl-2-methyl. The name to choose is the one that will give each group the lowest possible number.

P   The name of this compound is 5-ethyl-2-methylheptane.

2.   What is the IUPAC name for each isomer shown in the example in Section 12.3 of this study guide?

Isomer 1.

S   Apply the IUPAC nomenclature rules to Isomer 1

$$H-\underset{\underset{H}{|}}{\overset{\overset{H}{|}}{C}}-\underset{\underset{H}{|}}{\overset{\overset{H}{|}}{C}}-\underset{\underset{H}{|}}{\overset{\overset{H}{|}}{C}}-\underset{\underset{H}{|}}{\overset{\overset{H}{|}}{C}}-H$$

T•E   There are four carbon atoms in the main carbon chain, each connected by a single bond. Therefore, this compound is an alkane. The correct prefix is but- (4 carbons), and suffix -ane (alkane). There are no groups other than hydrogen attached to the carbon chain.

P   The IUPAC name of this compound is butane.

Isomer 2.

S   Once again, apply the IUPAC nomenclature rules.

$$H-\underset{\underset{H}{|}}{\overset{\overset{H}{|}}{C}}-\underset{\underset{H}{|}}{\overset{\overset{\overset{\overset{H}{|}}{C-H}}{|}}{C}}-\underset{\underset{H}{|}}{\overset{\overset{H}{|}}{C}}-H$$

T•E   There are three carbon atoms in the main carbon chain, and this is also an alkane. Therefore, the prefix will be prop-, and the suffix -ane. There is a group attached to the main carbon chain, and it is the methyl group. This group is attached to the second carbon.

P   The IUPAC name is 2-methylpropane.

146

10. Name the nine structural isomers of C_7H_{16} given in the answer to Check Your Understanding question 9.
11. Write the structural formula for each of the following compounds:
    (a) propane
    (b) octane
    (c) 3-ethylnonane
    (d) 3-methylheptane
    (e) 2,2-dichloropentane
    (f) 1,3-dibromopropane
    (g) 4-ethyl-2-methylhexane
    (h) 2,2,3,3-tetramethylbutane

12.7   Reactions of Alkanes

Although alkanes are fairly unreactive, they do undergo several chemical reactions, including oxidation. Our cells meet their energy needs through oxidation of compounds that are derivatives of alkanes. Oxidation is an exothermic reaction between a compound and oxygen. When the reaction rate is very fast and the heat produced can be felt and seen, we refer to oxidation as combustion. The products of complete oxidation of an alkane are carbon dioxide and water. If there is an insufficient supply of oxygen present when alkanes burn, incomplete oxidation (incomplete combustion) occurs and the products are carbon monoxide and water.

Alkanes also undergo substitution reactions with the halogens. In a substitution reaction, an atom or group of atoms replaces (is substituted for) a hydrogen on the alkane.

Important Terms

oxidation   substitution reaction   combustion   halogenation

For questions 12 to 19, fill in the blank with the correct word or words.
12. A major industrial source of hydrocarbons is _____, which was formed from _____.

13. The hydrocarbons in this source are separated from one another using a process called _____, which makes use of the fact that short-carbon-chain hydrocarbons have _____ boiling points than long-chain hydrocarbons.
14. Methane and ethane belong to a class of hydrocarbons called _____. The general molecular formula for this class of compounds is _____.
15. A hydrocarbon that contains all carbon-to-carbon single bonds is _____.
16. Compounds that have identical molecular formulas, but different arrangements of atoms in their molecules, are called _____.

17. Alkanes are relatively _____ (reactive, unreactive), but they do undergo _____ and _____ reactions.

18. The products of complete combustion (oxidation) of an alkane are _____ and the products of incomplete combustion (oxidation) are _____.

19. Complete and balance the equations for the following reactions:

(a) $CH_3CH_3 + O_2 \longrightarrow$

(b) $CH_3CH_3 + Br_2 \xrightarrow{\text{light}}$

12.8 Unsaturated Hydrocarbons

The unsaturated hydrocarbons contain carbon-to-carbon double bonds and carbon-to-carbon triple bonds. Compounds that contain more than one double or triple bond are called polyunsaturated. Double or triple bonds form an unstable spot in the hydrocarbon molecule. As a result, unsaturated hydrocarbons are more reactive than saturated hydrocarbons.

The alkenes are the class of hydrocarbons that contain one or more carbon-to-carbon double bonds. The general formula for alkenes containing one double bond is C_nH_{2n}. The alkynes are unsaturated hydrocarbons that contain carbon-to-carbon triple bonds. The general formula for alkynes containing one triple bond is C_nH_{2n-2}.

Important Terms

unsaturated   polyunsaturated   alkenes   alkynes

12.9 The Structure and Nomenclature of Alkenes

To name an alkene, you follow the same rules we discussed for the alkanes except that the name ends in -ene. The nomenclature rules have a series of priorities. The double bond takes precedence over the single bond and over alkyl substituents; therefore, the main carbon chain must be the one that contains the double bond. As was the case with substituent groups, the position of the double bond is indicated by a number. If there is more than one double bond, a prefix is used to indicate the number of double bonds.

Example _____

1. Name this compound

$$CH_3CH_2CHCH_2CH_3$$
$$|$$
$$CH_2CH=CH_2$$

148

S   Apply the IUPAC rules to name this compound.

T•E   (a)   The main carbon chain of an alkene must be the one containing the double bond. It will be easier to determine the name if we rewrite the structure as follows:

$$CH_2CH_3$$
$$|$$
$$CH_3CH_2CHCH_2CH=CH_2$$

Rewritten as shown, we see that the main carbon chain has six carbons: a hexene. The double bond is attached to the first carbon, so this is now 1-hexene.

(b)   There is one substituted group: an ethyl group (see Table 12.2 for the names of such groups). The substituted group is on carbon 4.

P   The name is 4-ethyl-1-hexene.

2. Write the structural formula for 3-ethyl-1,3-pentadiene.

S   Apply the IUPAC nomenclature rules to draw the structural formula.

T•E   (a)   The first step in writing a structural formula is to write the carbons in the main carbon chain. The prefix "penta-" tells us that there are five carbons.

C C C C C

(b)   The suffix "-diene" indicates that there are two double bonds, one on carbon 1 and the second on carbon 3. The rest of the bonds between the carbons are single bonds.

C=C-C=C-C

(c)   There is one substituent group: an ethyl group ($-CH_2CH_3$) on carbon 3.

$$CH_2CH_3$$
$$|$$
$$C=C-C=C-C$$

(d)   The rest of the bonds on each carbon atom will be attached to hydrogen atoms. When adding hydrogens, we must make sure that each carbon has four

149

bonds attached to it.

P   The structural formula is

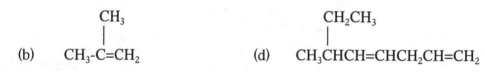

                                            or $CH_3-CH=\overset{\overset{\displaystyle CH_2CH_3}{|}}{C}-CH=CH_2$

Check Your Understanding _____

20.   Write the structural formula for the following compounds.
   (a)   propene   (c)   4-propyl-2,4,6-octatriene
   (b)   2,3-dimethyl-2-butene   (d)   4-ethyl-2-methyl-2-hexene

21.   Name the following compounds.
   (a) $CH_3CH=CHCH_3$ (c) $CH_3CH_2CH_2CH=CH_2$

   (b) $CH_3-\overset{\overset{\displaystyle CH_3}{|}}{C}=CH_2$ (d) $CH_3CHCH=CHCH_2CH=CH_2$ with $\overset{\displaystyle CH_2CH_3}{|}$

12.10 Bonding in the Alkenes
The double bond actually consists of two bonds: the sigma bond and the pi bond. The sigma bond is formed by the overlap of s, p, or hybrid orbitals along the bond axis. The pi bond is formed by the overlap of p orbitals above and below the bond axis. A triple bond consists of one sigma and two pi bonds.

Important Terms

   sigma bond   pi bond

12.11 Cis-Trans Isomerism
The double bond in an alkene forms a rigid spot on the molecule, permitting two possible

arrangements of atoms around the double bond. This type of isomerism is called geometirc isomerism or cis-trans isomerism. The cis-isomer has the specified atoms on the same side of the bond. The trans-isomer has the specified atoms on opposite sides of the bond. Review the criteria for determining the existence of cis-trans isomerism that are found in Section 12.11 in the text.

$$Cl\diagdown \quad \diagup Cl$$
$$C{=}C$$
$$H\diagup \quad \diagdown H$$

$$Cl\diagdown \quad \diagup H$$
$$C{=}C$$
$$H\diagup \quad \diagdown Cl$$

cis-1,2-dichlorothene   trans-1,2-dichloroethene

Important Terms

cis-trans isomerism   geomteric isomerism
cis-isomer   trans-isomer

Check Your Understanding ——————————————————————————

22.   Draw and label the cis- and trans-isomers for the following compounds.
       (a)   1,2-difluoropropene   (b)   2-hexene

12.12 Reactions of the Alkenes

Alkenes are more reactive than alkanes. They readily undergo oxidation, addition, and polymerization reactions. For our study of organic chemistry, we will modify our definition of oxidation to denote reactions in which one of the reactant molecules gains oxygen atoms or loses hydrogen atoms. A reduction reaction occurs with every oxidation reaction. The compound that is reduced will be the one that loses oxygen atoms or gains hydrogen atoms. Alkenes, like alkanes, undergo combustion to form carbon dioxide and water.

An addition reaction occurs when two atoms (or groups of atoms) react with doubly bonded carbon atoms, resulting in the formation of a carbon-to-carbon single bond with one of the reacting atoms (or groups of atoms) bonded to each carbon atom. Water, the halogens, hydrogen, and the hydrohalogens (such as HCl) can be added to the double bond. Hydrogenation (the addition of hydrogen to the double bond) is important in the industrial production of margarine.

Take a close look at the addition of bromine to ethene:

$$H-\overset{\displaystyle H}{\underset{\displaystyle |}{C}}=\overset{\displaystyle H}{\underset{\displaystyle |}{C}}-H \quad + \quad Br-Br \quad \longrightarrow \quad H-\overset{\displaystyle H}{\underset{\displaystyle Br}{C}}-\overset{\displaystyle H}{\underset{\displaystyle Br}{C}}-H$$

The single bond between the bromines has been broken and one of the two bonds of the double bond in ethene has been broken. Two new bonds have been formed. They are the carbon to bromine bonds in the product, 1,2-dibromoethane. Bromine has been "added" to the alkene.

The process of joining small molecules together to form very large molecules is called polymerization. The small repeating units are called monomers, and the large molecule they form is called a polymer. Polymers occur naturally in living organisms and they are also produced synthetically. They are the subject of much of Section IV of the textbook.

Important Terms

reduction	oxidation	addition reaction
hydrogenation	hydration	bromination
polymerization	monomer	polymer

Check Your Understanding _____

For questions 23 to 28, choose the best answer or answers.

23. Alkenes are
    (a) more reactive than alkanes   (c) unreactive
    (b) less reactive than alkanes   (d) saturated

24. The addition of water to the double bond of an alkene is called a
    (a) halogenation   (c) hydration reaction
    (b) reduction reaction   (d) hydrogenation reaction

25. The production of margarine involves the addition of this compound to double bonds in vegetable oils.
    (a) water   (c) hydrogen chloride
    (b) hydrogen   (d) oxygen

26. When a reactant molecule gains oxygen atoms, the reaction that has occurred is a
    (a) oxidation reaction   (c) hydrogenation reaction
    (b) reduction reaction   (d) addition reaction

27. When monomer units are joined together to form a very large molecule, the reaction is a

   (a) addition reaction   (c) hydrogenation reaction
   (b) hydration reaction   (d) polymerization reaction

28. When a reactant molecule gains hydrogen atoms or loses oxygen atoms, the reaction that has occurred is a

   (a) oxidation reaction   (c) hydrogenation reaction
   (b) reduction reaction   (d) addition reaction

29. Draw the structure of the products of the following reactions.

   (a) $CH_3-CH_2-CH=CH_2 \ + \ Cl_2 \longrightarrow$

   (b) $CH_3-CH_2-CH=CH_2 \ + \ H_2 \longrightarrow$

   (c) $CH_3-CH_2-CH=CH_2 \ + \ H_2O \longrightarrow$

   (d) $CH_3-CH_2-CH=CH_2 \ \xrightarrow{\ \ KMnO_4 \ \ }$

   (e) The polymerization of three molecules of 1-butene

12.13 Alkynes

The alkynes are unsaturated hydrocarbons that contain the carbon-to-carbon triple bond. Alkynes are named following the same rules as the alkenes, except that the name ends in -yne.

Important Term

   alkyne

12.14 Cycloalkanes and Cycloalkenes

Carbon atoms can form rings containing single, double, and (in very large rings) triple bonds. The most common cyclic hydrocarbons contain five or six carbon atoms. This section also discusses the anesthetic properties of cyclic hydrocarbons such as cyclopropane.

There are several ways to draw the structures of cyclic compounds. Here we see the extended structural formula, the condensed structural formula and the line bond formula for cyclohexane. The line bond formula is the easiest to draw and is used most often. Remember that each point in the structure corresponds to a carbon and that each carbon must have four bonds. Any bonds that are not shown are to hydrogens.

The rotation of the carbon atoms around the single bonds in a cyclic hydrocarbon is restricted. This allows the formation of cis- and trans-isomers. The cis-isomer of a cyclic compound has the specified atoms attached on the same side of the carbon ring. The trans-isomer has the attached atoms on opposite sides of the ring.

cis-1,2-dichlorocyclopentane   trans-1,2-dichlorocyclopentane

Important Terms

    cyclic hydrocarbon   anesthetic

Check Your Understanding —————————————————————————

30.   Draw the structural formula for each of the following compounds.
      (a)   cyclopentene
      (b)   1,3,5-trichlorocyclohexane
      (c)   trans-1,2-dimethylcyclobutane
      (d)   1,3-diethylcyclopentane

12.15 Aromatic Hydrocarbons

The aromatic hydrocarbons are a class of cyclic hydrocarbons made up of benzene and its derivatives. Benzene, with a molecular formula of C_6H_6, has an unusual structure, often described as a "resonance hybrid". This structure makes the molecule very stable. Because of its stability, benzene and its derivatives form part of many complex naturally occurring compounds. Be sure to review the schematic ways of writing the structural formulas of cyclic and aromatic hydrocarbons presented in this section.

The structure of benzene makes the ring very stable and resistant to most chemical changes. The double bonds in the ring, unlike the double bonds in alkenes, do not undergo addition reactions. Benzene molecules do undergo reactions that involve the substitution of an atom or group of atoms for the hydrogens attached to the ring.

Important Terms

      benzene   resonance hybrid
      aromatic hydrocarbon   aliphatic hydrocarbon

Check Your Understanding —————————————————————————

For questions 31 to 34, answer true (T) or false (F).

31.   All the carbon-to-carbon bonds in benzene are equal.

32.   The double bonds in a benzene ring are less reactive than the double bonds in other alkenes.

33.   Animals can synthesize benzene rings.

34.   Benzene will readily undergo addition and substitution reactions.

35.   Draw the complete structural formulas represented by the following schematic formulas:

(a) benzene ring with NO_2 substituent

(b) cyclohexane ring with OH substituent

CH₃ ... (represented below)

(c) 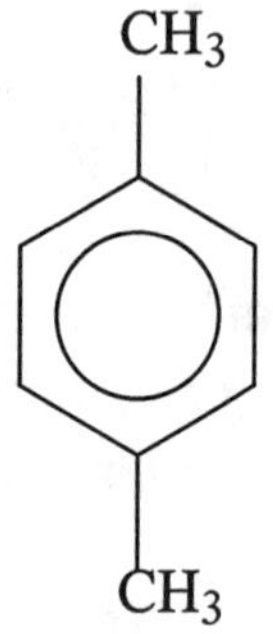 [benzene ring with CH$_3$ at top and CH$_3$ at bottom]

(d) 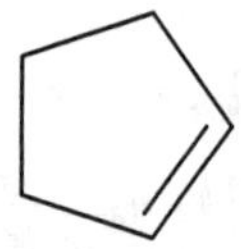 [cyclopentene ring]

36.   Write the equation for the reaction between benzene and chlorine.

Answers to Check Your Understanding Questions in Chapter 12

1. $CH_3CH_2CH_2CH_2CH_2CH_3$ **2.** $CH_3C(CH_3)_2CH_2CH_3$ **3.** $CH_3CH(CH_3)CH(CH_3)CH_3$
4. $CH_3CH(CH_3)CH=CH_2$

(5) 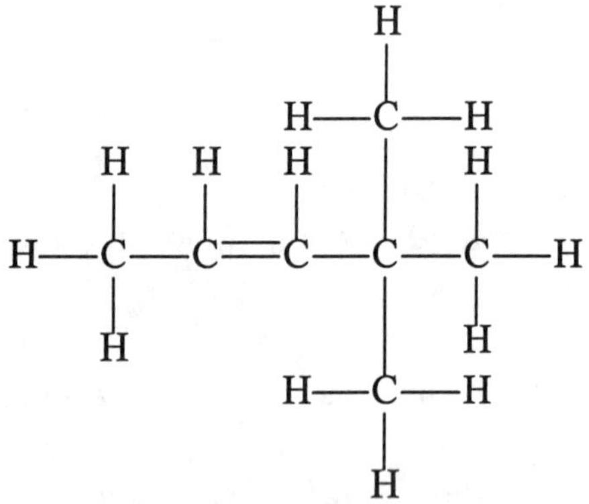
```
              H
              |
         H—C—H
  H   H   H     H
  |   |   |     |
H—C—C=C—C—C—H
  |       |
  H       H
       H—C—H
          |
          H
```

(6) 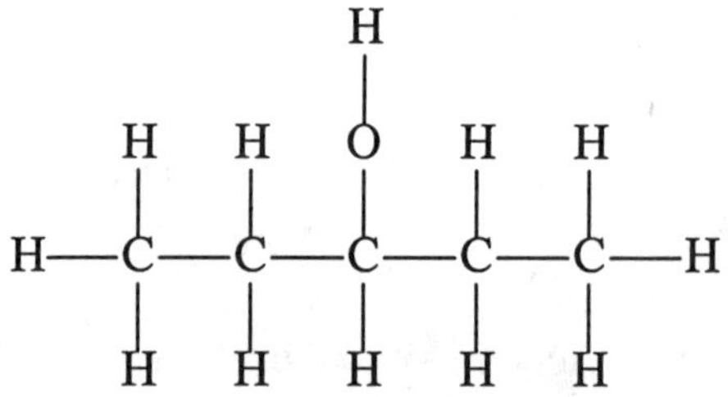
```
                  H
                  |
  H   H   O   H   H
  |   |   |   |   |
H—C—C—C—C—C—H
  |   |   |   |   |
  H   H   H   H   H
```

(7) 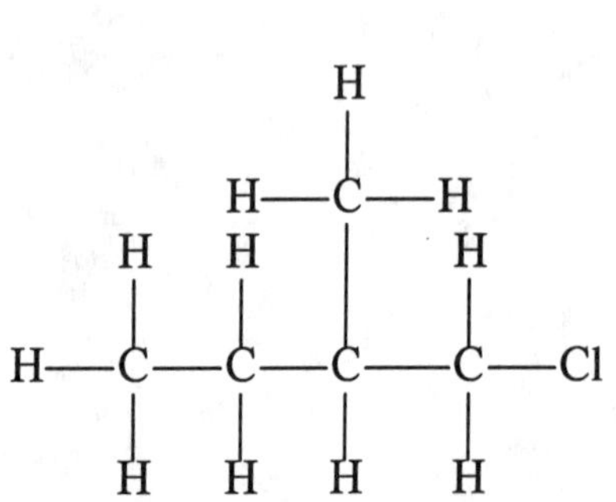
```
          H
          |
     H—C—H
  H   H   H
  |   |   |
H—C—C—C—C—Cl
  |   |   |   |
  H   H   H   H
```

(8) 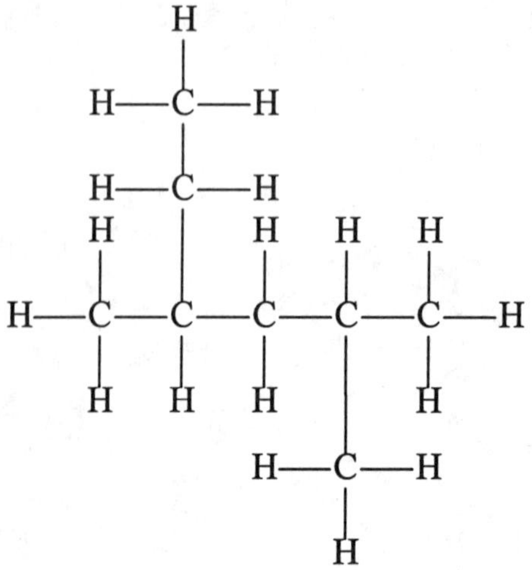
```
            H
            |
       H—C—H
       |
       H—C—H
       H     H   H   H
       |     |   |   |
  H—C—C—C—C—C—H
       |   |   |     |
       H   H   H     H
                 H—C—H
                    |
                    H
```

156

9.

10. (1) heptane (2) 2-methylhexane (3) 3-methylhexane (4) 2,3-dimethylpentane (5) 2,4-dimethylpentane (6) 2,2-dimethylpentane (7) 3,3-dimethylpentane (8) 2,2,3-trimethylbutane (9) 3-ethylpentane

11. (a) $CH_3CH_2CH_3$ (b) $CH_3CH_2CH_2CH_2CH_2CH_2CH_2CH_3$

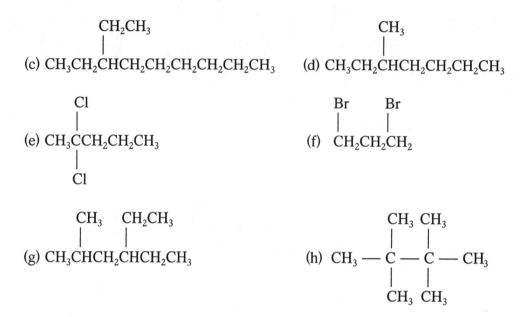

(c) $CH_3CH_2\overset{\overset{\displaystyle CH_2CH_3}{|}}{C}HCH_2CH_2CH_2CH_2CH_2CH_3$ (d) $CH_3CH_2\overset{\overset{\displaystyle CH_3}{|}}{C}HCH_2CH_2CH_2CH_3$

(e) $CH_3\underset{\underset{\displaystyle Cl}{|}}{\overset{\overset{\displaystyle Cl}{|}}{C}}CH_2CH_2CH_3$ (f) $\overset{\overset{\displaystyle Br}{|}}{C}H_2\overset{\overset{\displaystyle Br}{|}}{C}H_2CH_2$

(g) $CH_3\overset{\overset{\displaystyle CH_3}{|}}{C}HCH_2\overset{\overset{\displaystyle CH_2CH_3}{|}}{C}HCH_2CH_3$ (h) $CH_3 - \overset{\overset{\displaystyle CH_3}{|}}{\underset{\underset{\displaystyle CH_3}{|}}{C}} - \overset{\overset{\displaystyle CH_3}{|}}{\underset{\underset{\displaystyle CH_3}{|}}{C}} - CH_3$

12. petroleum, decayed plant material **13.** fractional distillation, lower **14.** alkanes, C_nH_{2n+2}
15. saturated **16.** constitutional isomers **17.** unreactive, oxidation and substitution **18.** CO_2 and H_2O, CO and H_2O

19. (a) $2\,CH_3CH_3 + 7\,O_2 \longrightarrow 4\,CO_2 + 6\,H_2O$

(b) $CH_3CH_3 + Br_2 \longrightarrow CH_3CH_2Br + HBr$

20. (a) $CH_3CH=CH_2$ (b) $CH_3-\overset{\overset{\displaystyle CH_3}{|}}{C}=\overset{\overset{\displaystyle CH_3}{|}}{C}-CH_3$ (c) $CH_3\text{-}CH=CH\text{-}\overset{\overset{\displaystyle CH_2CH_2CH_3}{|}}{C}=CH\text{-}CH=CH\text{-}CH_3$

(d) $CH_3\overset{\overset{\displaystyle CH_3}{|}}{C}=CHCH\underset{\underset{\displaystyle CH_2CH_3}{|}}{C}H_2CH_3$

21. (a) 2-butene
(b) 2-methyl-1-propene
(c) 1-pentene
(d) 6-methyl-1,4-octadiene
(check the longest carbon chain)

22.

(a)

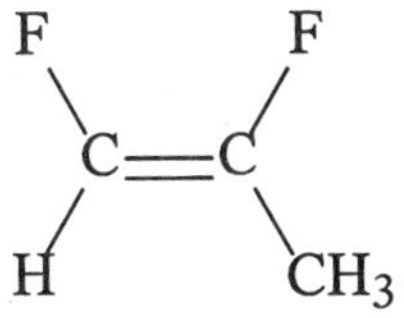

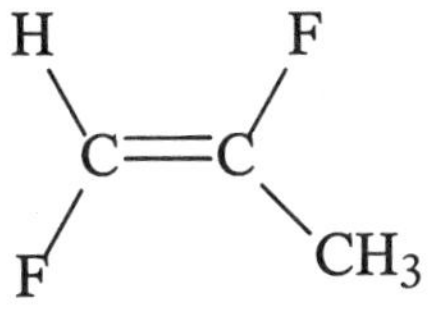

cis

$$\begin{array}{ccc} F & & F \\ | & & | \\ C & = & C \\ | & & | \\ H & & CH_3 \end{array}$$

trans

$$\begin{array}{ccc} H & & F \\ | & & | \\ C & = & C \\ | & & | \\ F & & CH_3 \end{array}$$

(b)

$$\begin{array}{ccc} CH_3 & & CH_2CH_2CH_3 \\ | & & | \\ C & = & C \\ | & & | \\ H & & H \end{array}$$

$$\begin{array}{ccc} H & & CH_2CH_2CH_3 \\ | & & | \\ C & = & C \\ | & & | \\ CH_3 & & H \end{array}$$

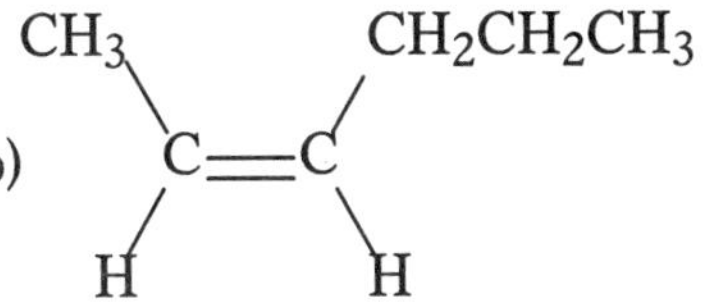

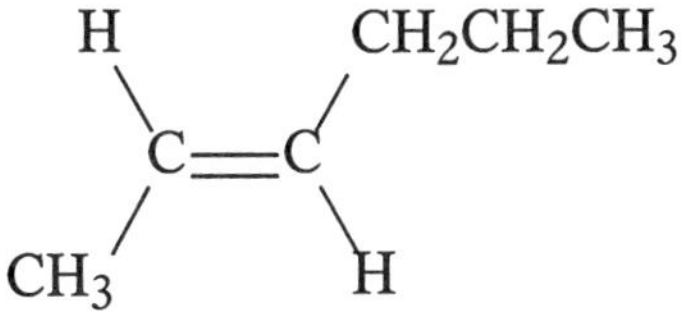

23. a **24.** c **25.** b **26.** a **27.** d **28.** b

29.

(a)

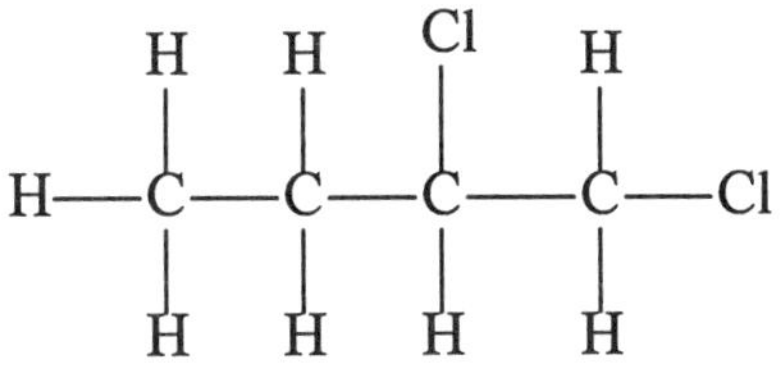

$$\begin{array}{ccccc} & H & H & Cl & H \\ & | & | & | & | \\ H - & C - & C - & C - & C - Cl \\ & | & | & | & | \\ & H & H & H & H \end{array}$$

(c)

$$\begin{array}{ccccc} & & & H & \\ & & & | & \\ & H & H & O & H \\ & | & | & | & | \\ H - & C - & C - & C - & C - H \\ & | & | & | & | \\ & H & H & H & H \end{array}$$

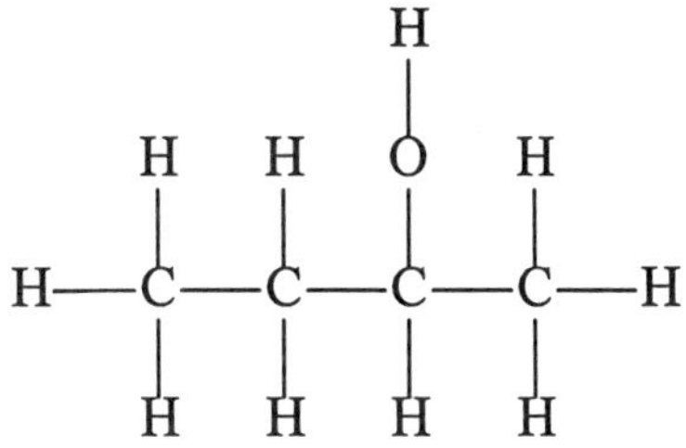

(b)

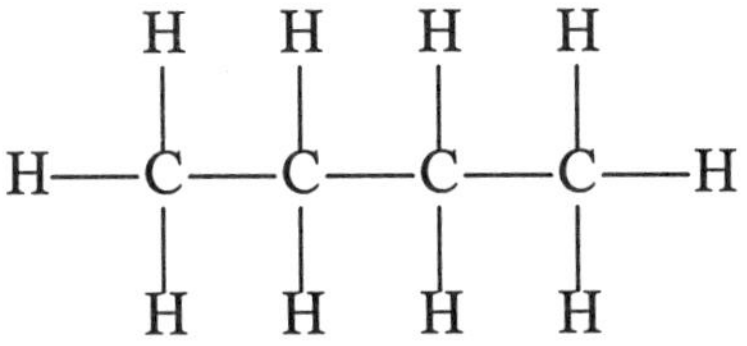

$$\begin{array}{ccccc} & H & H & H & H \\ & | & | & | & | \\ H - & C - & C - & C - & C - H \\ & | & | & | & | \\ & H & H & H & H \end{array}$$

(d)

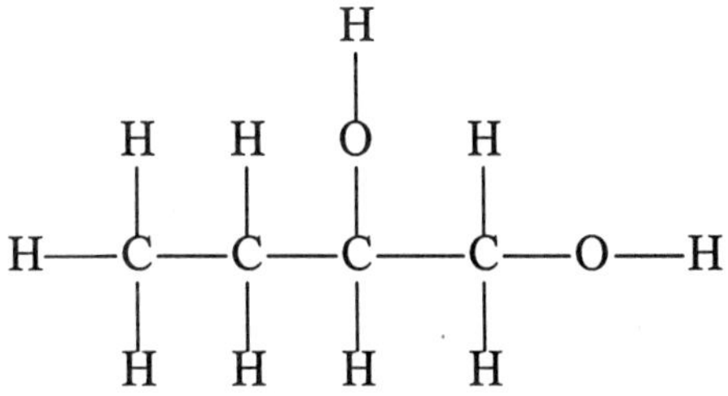

$$\begin{array}{ccccc} & & & H & \\ & & & | & \\ & H & H & O & H \\ & | & | & | & | \\ H - & C - & C - & C - & C - O - H \\ & | & | & | & | \\ & H & H & H & H \end{array}$$

(e)

$$\begin{array}{cccccc} & H & C_2H_5 & H & C_2H_5 & H & C_2H_5 \\ & | & | & | & | & | & | \\ H - & C - & C - & C - & C - & C - & C - H \\ & | & | & | & | & | & | \\ & H & H & H & H & H & H \end{array}$$

159

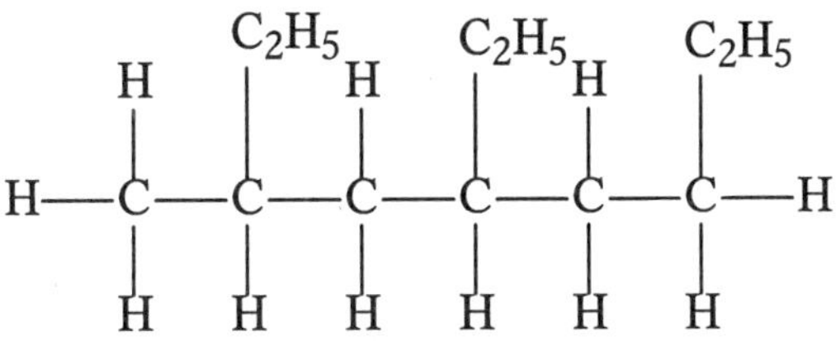

30.

(a)

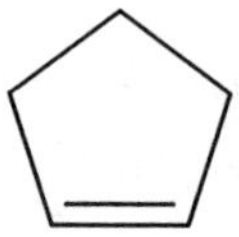

(b)

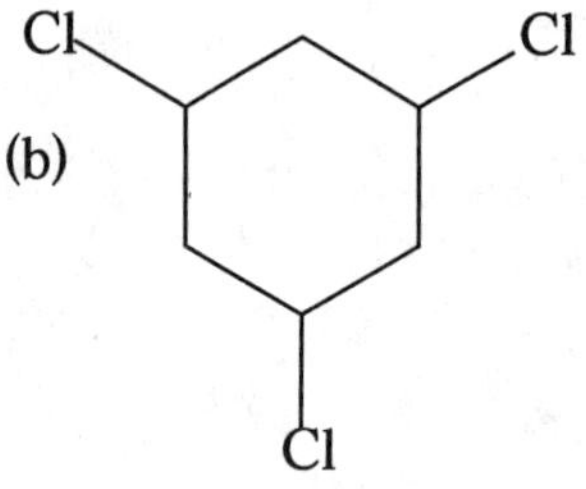

(c)

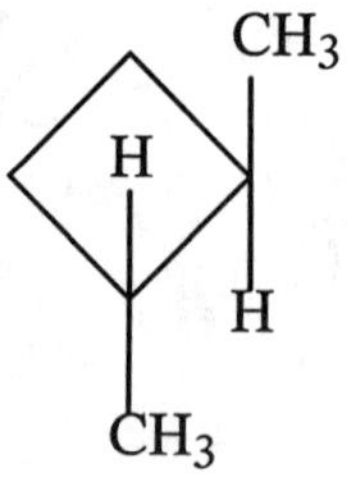

(d)

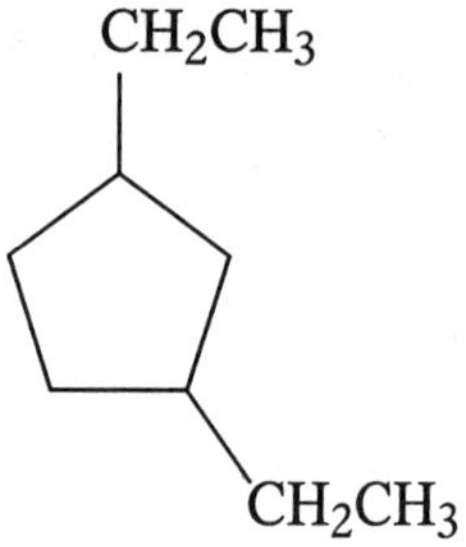

31. T **32.** T **33.** F **34.** F

35.

(a)

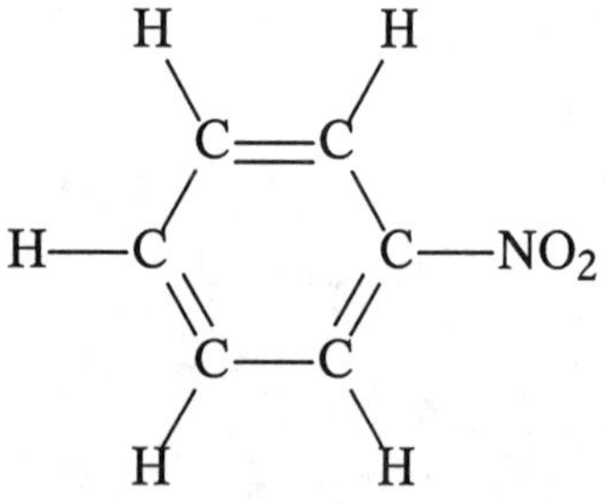

(b)

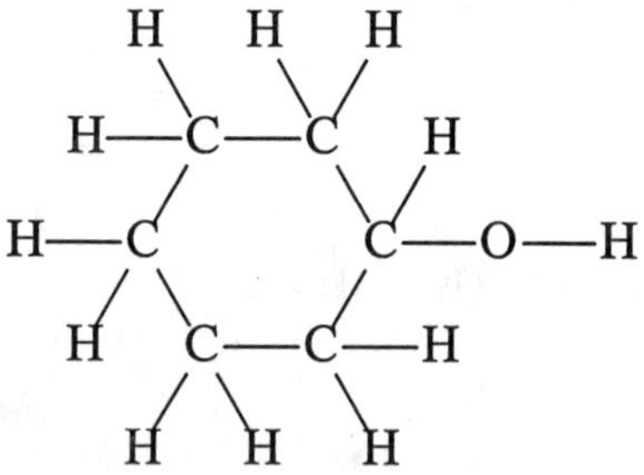

(c)

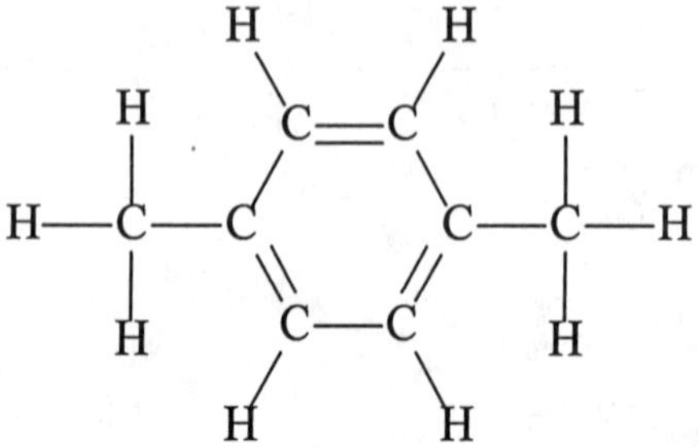

(d)

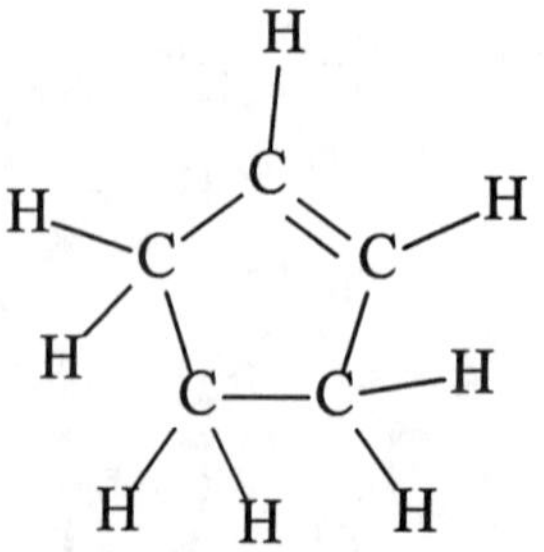

36.

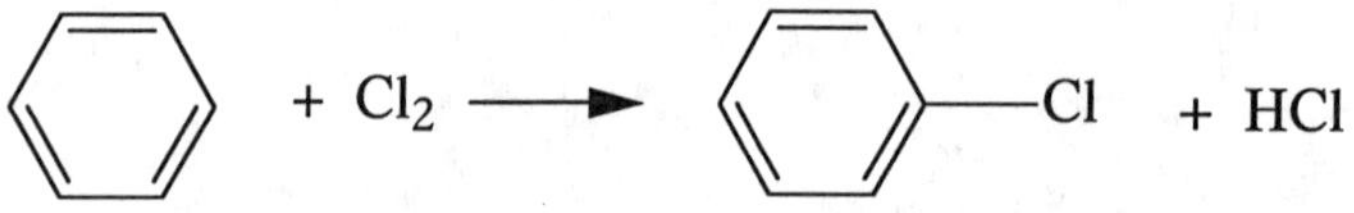

160

Chapter 13   ORGANIC COMPOUNDS
CONTAINING OXYGEN

The wide variety of hydrocarbons described in the previous chapters only hints at the full range of organic compounds. Many more types of organic compounds are formed when elements such as oxygen, nitrogen, phosphorus, and sulfur combine with hydrocarbons. In this chapter we study organic compounds that contain oxygen. These compounds are classified based on their oxygen-containing functional groups.

13.1   Functional Groups Containing Oxygen

Functional groups are arrangements of atoms that create a reactive area on an organic molecule and that give the molecule specific chemical properties. In this chapter, we study the functional groups containing oxygen. There are numerous oxygen-containing functional groups, that might look very similar to each other at first. Learn to recognize these functional groups.

The letter "R" is used again in this chapter to refer to an alkyl group such as the ethyl group or the methyl group. Use of the "R" symbol helps to simplify structural formulas.

13.2   Ethers

The oxygen in an ether molecule is bonded to two carbon atoms, -C-O-C-. The general formula is R-O-R'. The two alkyl groups can be the same or they can be different. Simple ethers are named by listing both alkyl groups attached to the oxygen. An ether molecule is nonpolar; it is soluble in nonpolar solvents and not soluble in water. Ethers are highly flammable, and many are used as anesthetics.

Important Terms

      ether   anesthetic

Check Your Understanding

1. Write the structural formulas for the following compounds:
   (a)   diethyl ether   (b)   methyl propyl ether
2. Name the following compounds:
   (a) $CH_3CH_2CH_2CH_2OCH_2CH_3$ (b) $CH_3\text{-}CH\text{-}O\text{-}CH\text{-}CH_3$
   $$||$$
   $$CH_3CH_3$$

13.3 Alcohols

The functional group that makes alcohols a distinct class of compounds is the hydroxyl group, -OH. Specific alcohols are named by changing the -e ending of the name of the parent alkane to -ol. The position of the hydroxyl group in the molecule is indicated by placing a number directly in front of the name of the parent compound. Thus, the name of this alcohol is 2-butanol.

$$H-\overset{\overset{\displaystyle H}{|}}{\underset{\underset{\displaystyle H}{|}}{C}}-\overset{\overset{\displaystyle H}{|}}{\underset{\underset{\displaystyle H}{|}}{C}}-\overset{\overset{\displaystyle O-H}{|}}{\underset{\underset{\displaystyle H}{|}}{C}}-\overset{\overset{\displaystyle H}{|}}{\underset{\underset{\displaystyle H}{|}}{C}}-H$$

Alcohols are classified based on the position of the hydroxyl group in the molecule into primary, secondary, or tertiary alcohols. Primary alcohols have one carbon and two hydrogens bonded to the carbon to which the hydroxyl is bonded. Secondary alcohols have two carbons and one hydrogen on the carbon bonded to the hydroxyl and tertiary alcohols have three carbons and no hydrogens on the carbon bonded to the hydroxyl group.

$$H-\overset{\overset{\displaystyle H}{|}}{\underset{\underset{\displaystyle OH}{|}}{C}}-R \qquad H-\overset{\overset{\displaystyle R}{|}}{\underset{\underset{\displaystyle OH}{|}}{C}}-R \qquad R-\overset{\overset{\displaystyle R}{|}}{\underset{\underset{\displaystyle OH}{|}}{C}}-R$$

primary	secondary	tertiary

The hydroxyl group creates a reactive polar area on the unreactive, nonpolar hydrocarbon. Short-carbon-chain alcohols are soluble in water because of hydrogen bonding that forms between the hydroxyl group and the water molecules. Hydrogen bonds can also form between alcohol molecules, causing alcohols to have higher melting and boiling points than alkanes with comparable molecular weights.

Important Terms

| alcohol | hydroxyl group | |
| primary alcohol | secondary alcohol | tertiary alcohol |

1. Name the following compound:

$$OH$$
$$|$$
$$CH_3\text{-}CH_2\text{-}CH\text{-}CH\text{-}CH_2\text{-}CH_3$$
$$|$$
$$CH_2CH_3$$

S   Apply the IUPAC nomenclature rules to name this compound.

T•E   The number of carbon atoms in the longest carbon chain to which the hydroxyl group is attached is 6. Therefore, this is a hexanol.

The hydroxyl group is on carbon 3 (counting from the right so that the hydroxyl group will have the lowest number). Therefore this is a 3-hexanol.

The other substituent group is an ethyl group. Based on the numbering of the carbon chain based on the position of the hydroxyl group, this group is attached to carbon 4.

P   The complete name for this alcohol is 4-ethyl-3-hexanol.

2. Write the structural formula for 4-chloro-1-pentanol.

S   Apply the nomenclature rules to determine the structural formula.

T•E   The name pentanol indicates that the main carbon chain has five carbons.

$$C\text{-}C\text{-}C\text{-}C\text{-}C$$

The hydroxyl group is attached to carbon 1 and a chlorine atom to carbon 4.

P   The structure is:

$$Cl$$
$$|$$
$$CH_3\text{-}CH\text{-}CH_2\text{-}CH_2\text{-}CH_2\text{-}OH$$

3. Write the structural formulas for the following compounds:
    (a)   1-propanol   (b)   2-propanol

(c)   2,3-dimethyl-2-butanol   (d)   3,4-dibromo-2-hexanol

4.   Name the following compounds:

           OH
           |
(a) CH_3-CH-CH_2-CH_3 (c) $CH_3CH_2CH_2CH_2CH_2OH$

           CH_3 H
           | |
(b) CH_3-CH_2-C-CH_3 (d) CH_3-CH — C — H
           | | |
           OH CH_3 OH

5.   Indicate whether each of the compounds shown in question 4 is a primary, secondary, or tertiary alcohol.

6.   List the following in order of decreasing solubility in water: hexanol, ethanol, butanol.

13.4   Some Important Alcohols

You are probably most familiar with ethanol and methanol and you should know their structures. Polyhydric alcohols are alcohols that have more than one hydroxyl group. Important examples are ethylene glycol (1,2-ethanediol) and glycerol (1,2,3-propanediol).

Cyclic alcohols have one or more hydroxyl groups substituted for hydrogens on the parent cyclic hydrocarbon. When hydroxyl groups replace one or more hydrogens on a benzene ring, the class of compounds known as the phenols is formed. The compound with the name phenol is formed when just one hydrogen is replaced on the benzene ring. It is the simplest phenol and gives its name to the whole class of compounds. Do not confuse the phenols with the alcohols. The phenols do not undergo the same reactions as the alcohols. Phenols are not secondary alcohols.

Important Terms

      polyhydric alcohol   cyclic alcohol   phenol

13.5   Synthesis of Alcohols

Alcohols can be prepared by the addition of water to an alkene (hydration reaction), or by the reduction of an aldehyde or ketone.

13.6   Reactions of Alcohols

Alkenes can be produced by the dehydration of an alcohol. Oxidation (dehydrogenation) of primary alcohols, under controlled conditions, produces aldehydes. Under more vigorous conditions, primary alcohols are oxidized to carboxylic acids. Oxidation of secondary alcohols produces ketones. Tertiary alcohols are not easily oxidized.

Example
_____

Complete the following reaction:

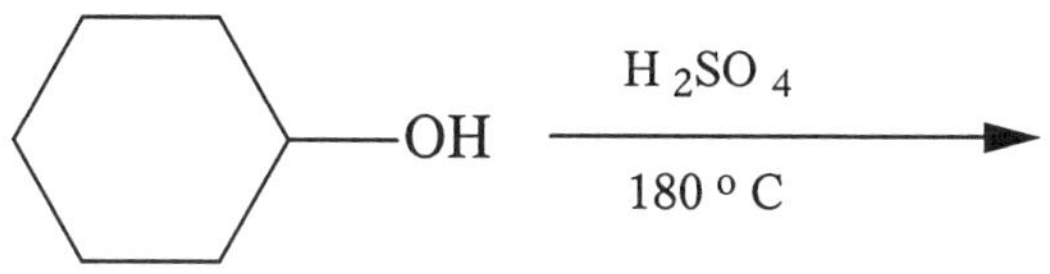

S   We must decide what the products of this reation are and then draw their structures.

T•E   An alcohol is being treated with sulfuric acid at high temperature. Under these conditions the alcohol undergoes dehydration. The hydroxyl group and a hydrogen on an adjacent carbon must be removed and these two carbons must be joined by a double bond. In addition to an alkene, the dehydration of an alcohol produces water.

P   The complete reaction is:

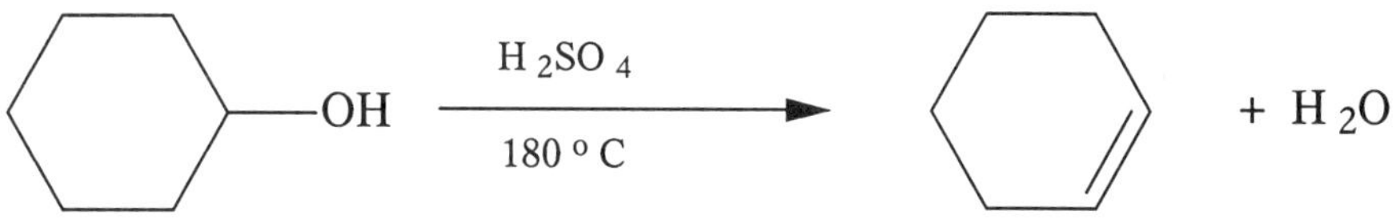

Check Your Understanding   _____

For questions 7 to 13, fill in the blank with the correct word or words.

7.   The formula of wood alcohol is _____ .
8.   The end products of fermentation are _____ .
9.   Ethylene glycol is a _____ alcohol. _____ is a trihydric alcohol that is found in body fat.
10.   The compound containing one hydroxyl group on a benzene ring is called _____ .
11.   Alcohols are produced by the _____ of an aldehyde or ketone, or by the _____ of an alkene.

12. The dehydration of propanol produces _____.

13. Aldehydes are produced by the oxidation of _____ alcohols, and ketones by the oxidation of _____ alcohols.

14. Write the formula for the product of the dehydration of each of the following alcohols.

(a) $CH_3-CH_2-CH_2-CH_2-OH$

(c) $-CH_2-CH_2-OH$

(b)
$$CH_3-\underset{\underset{OH}{|}}{CH}-CH_3$$

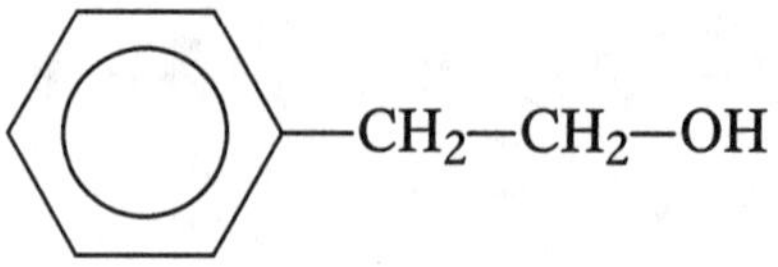

(d)
$$CH_3-\underset{\underset{CH_3}{|}}{CH}-CH_2-OH$$

15. Write the formula for the product of the oxidation of each of the following alcohols.

(a) $CH_3CH_2CH_2OH$

(c)
$$CH_3-CH_2-\underset{\underset{CH_3}{|}}{CH}-OH$$

(b)
$$CH_3-\underset{\underset{CH_3}{|}}{CH}-\underset{\underset{OH}{|}}{CH}-CH_3$$

(d)
$$CH_3CH_2\underset{\underset{CH_3}{|}}{CH}CHOH \quad \underset{CH_3}{}$$

13.7 Aldehydes and Ketones

Both aldehydes and ketones contain the carbonyl functional group. Aldehydes have the carbonyl group on the terminal carbon of a chain. Ketones have the carbonyl group on a carbon that is not at the end of a chain.

Learn to recognize the difference in structure between aldehydes and ketones. In an aldehyde, the carbonyl carbon has one or two hydrogens attached to it. In a ketone there are two carbons but no hydrogens attached to the carbonyl carbon.

aldehyde   ketone

Aldehydes are named by adding an -al ending to the name of the longest carbon chain containing the carbonyl group. The names of ketones have an -one ending, and use a number before the name to indicating the position of the carbonyl group.

Important Terms

carbonyl group (-C-)   aldehyde   ketone

13.8   Some Important Aldehydes and Ketones
Many aldehydes and ketones have pleasant odors and tastes, and are used in making perfumes and flavorings. Formaldehyde is a well-known aldehyde, and acetone is a widely used ketone.

13.9   Synthesis of Aldehydes and Ketones
Aldehydes are produced by the controlled oxidation of primary alcohols. Ketones are produced by the oxidation of secondary alcohols.

13.10 Reactions of Aldehydes and Ketones

Reduction of aldehydes and ketones is carried out in the presence of a catalyst such as platinum. The reduction of an aldehyde produces a primary alcohol, and the reduction of a ketone a secondary alcohol. Aldehydes are easily oxidized to form carboxylic acids, but ketones can be oxidized only under extreme chemical conditions.

When an aldehyde reacts with an alcohol, a hemiacetal is formed. A similar molecule, called a hemiketal, is formed when a ketone reacts with an alcohol. Cyclic hemiacetals or hemiketals can form when an aldehyde or ketone group reacts with an alcohol group on the same molecule. When you study carbohydrates, you will learn about the importance of these hemiacetals, hemiketals, acetals, and ketals in living organisms.

<pre>
          OH   OCH₃
          | |
CH₃——C*—OCH₃   CH₃——C*—OCH₃
          | |
          H   H

      hemiacetal   acetal
</pre>

This hemiacetal is the product of the reaction between methanol and ethanal. The acetal is the product of the reaction of the hemiacetal with a second molecule of methanol. Notice the carbons with the asterisk in these two examples. In the hemiacetal this carbon is bonded to a hydrogen, a hydroxyl group and an OCH_3 group. The hemiacetal can be recognized because it is an alcohol and an ether at the same carbon. In the acetal, the asterisked carbon is bonded to a hydrogen and two OCH_3 groups. The acetal can be recognized because it has two ether groups on the same carbon.

Important Terms

        hydrogenation   hemiacetal   hemiketal

Example   _____

Write the structure of the hemiacetal formed when propanal reacts with ethanol.

S   Draw the structures of propanal and ethanol and then draw the structure of the hemiacetal formed when they react.

T•E   The structures of propanal and ethanol are:

168

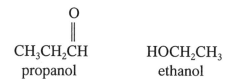

CH₃CH₂CH   HOCH₂CH₃
propanol   ethanol

The formation of a hemiacetal can be seen as the addition of the alcohol across the carbonyl bond. The H from the alcohol bonds to the oxygen of the carbonyl and the -OR of the alcohol bonds to the carbon of the carbonyl.

P   The structure of the hemiacetal is:

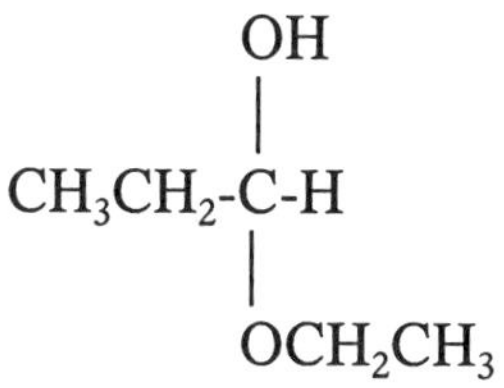

$$OH$$
$$|$$
$$CH_3CH_2\text{-}C\text{-}H$$
$$|$$
$$OCH_2CH_3$$

Check Your Understanding _____

16.   Name the following compounds, and indicate whether the compound is an aldehyde or a ketone.

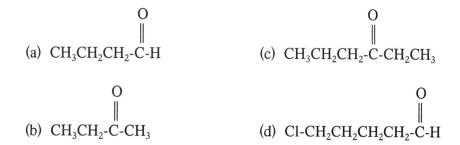

(a) CH₃CH₂CH₂-C-H (with O double bonded to C)

(c) CH₃CH₂CH₂-C-CH₂CH₃ (with O double bonded to C)

(b) CH₃CH₂-C-CH₃ (with O double bonded to C)

(d) Cl-CH₂CH₂CH₂CH₂-C-H (with O double bonded to C)

17.   Write the structural formula of the main product of each of the following reactions:

(a) 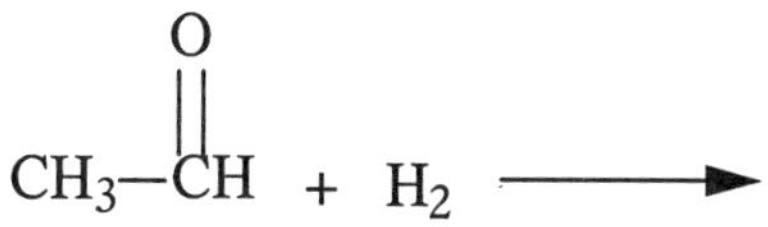   CH₃—CH + H₂ ⟶ (with O double bonded to CH)

(b)
$$CH_3-CH_2-CH_2-\overset{\overset{\displaystyle O}{\|}}{C}H \xrightarrow{\text{KMnO}_4}$$

(c)
$$CH_3-CH_2-\overset{\overset{\displaystyle OH}{|}}{C}H-CH_3 \xrightarrow{\text{KMnO}_4}$$

(d)
$$CH_3-CH_2-\overset{\overset{\displaystyle O}{\|}}{C}-CH_2-CH_3 \quad + \quad H_2 \xrightarrow{\text{Pt}}$$

(e)
$$CH_3-CH_2-CH_2-\overset{\overset{\displaystyle O}{\|}}{C}H \quad + \quad CH_3OH \longrightarrow$$

(f)
$$CH_3-\overset{\overset{\displaystyle O}{\|}}{C}-CH_3 \quad + \quad CH_3CH_2OH \longrightarrow$$

(g)
$$CH_3-\overset{\overset{\displaystyle CH_3}{|}}{C}H-CH_2-CH_2-OH \xrightarrow{\text{KMnO}_4}$$

13.11 Carboxylic Acids

Carboxylic acids are organic acids containing the carboxyl or carboxylic acid functional group. Carboxylic acids are weak acids. They are named by adding the suffix -oic acid to the name of the longest carbon chain containing the carboxyl group. Because the carboxyl group is polar, hydrogen bonds can form between carboxylic acid molecules, and between carboxylic acid molecules and water. Carboxylic acids have higher boiling points than comparable alcohols and are more soluble in water.

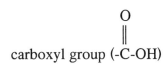

carboxylic acid   carboxyl group (-C-OH)

Check Your Understanding ──

18.   Name the following compounds:

$$O$$
$$\|$$
(a) $CH_3CH_2CH_2CH_2C\text{-}OH$

$$CH_3 \quad\quad O$$
$$| \quad\quad\quad \|$$
(b) $CH_3CHCHCH_2CH_2C\text{-}OH$
$$|$$
$$CH_3$$

19.   Write the structural formula for each of the these compounds:
        (a) propanoic acid   (b) 3-iodobutanoic acid

13.12 Some Important Carboxylic Acids

Formic acid, which causes the irritation of insect bites, is also produced in the liver during the oxidation of methanol. In sufficient quantities, formic acid can cause a severe metabolic disturbance called acidosis. Such acidosis is the cause of death from methanol poisoning. Acetic acid gives the characteristic taste to vinegar, and is used by cells to synthesize fatty acids. Other important carboxylic acids mentioned in this section are benzoic acid and oxalic acid.

13.13 Synthesis of Carboxylic Acids

Carboxylic acids are prepared by oxidizing primary alcohols or aldehydes with strong oxidizing agents.

Check Your Understanding ──

For questions 20 to 28, select the best answer(s).

20.   The functional group found in carboxylic acids is
        (a)   carbonyl group   (c)   hydroxyl group
        (b)   carboxyl group   (d)   amide group
21.   When placed in water, propionic acid forms
        (a)   the propionate ion   (c)   the hydroxide ion
        (b)   hydrogen propionate   (d)   the hydronium ion
22.   The strength of organic acids are comparable to the strength of
        (a)   nitric acid   (c)   sulfuric acid
        (b)   carbonic acid   (d)   hydrochloric acid

23. The stings of bees contain the carboxylic acid called
    (a) butyric acid   (c) formic acid
    (b) acetic acid   (d) lactic acid
24. Methanol is first oxidized in the liver to this compound which can cause blindness.
    (a) formic acid   (c) acetone
    (b) formaldehyde   (d) acetic acid
25. Vinegar contains
    (a) acetic acid   (c) butyric acid
    (b) formic acid   (d) lactic acid
26. The sodium salt of this acid is a commonly used food preservative.
    (a) lactic acid   (c) capric acid
    (b) tartaric acid   (d) benzoic acid
27. Oxidation of 1-propanol by $KMnO_4$ will produce
    (a) propanal   (c) propanone
    (b) propanoic acid   (d) propanoate
28. Oxidation of butanal by $KMnO_4$ will produce
    (a) butanoic acid   (c) 1-butanol
    (b) butanone   (d) pentanoic acid

13.14 Reactions of Carboxylic Acids

Carboxylic acids are proton donors. They do not donate a proton as readily as acids such inorganic acids as hydrochloric acid, so they are considered to be weak acids. Another important reaction of carboxylic acids is their reaction with alcohols to form esters.

Important terms

       esterification   condensation

13.15 Esters

Esters contain the ester functional group.

$$R-\overset{\displaystyle O}{\overset{\displaystyle \|}{C}}-OR'$$

Do not confuse esters and ethers. Although the names of these families are similar, esters and ethers have very different chemical properties. Look closely at the two examples below to see the structural difference between esters and ethers.

$$CH_3-\overset{\displaystyle O}{\overset{\displaystyle \|}{C}}-O-CH_2-CH_3 \qquad CH_3-CH_2-O-CH_2-CH_3$$

       ethyl acetate   diethyl ether

Esters are produced by the esterification (condensation) reaction between an alcohol and an acid. Esters have distinctive flavors and tastes. Esters are named by first listing the alcohol name with a -yl ending, and then the acid name with an -ate ending.

$$CH_3-\overset{\overset{\displaystyle O}{\|}}{C}-OH \ + \ CH_3CH_2CH_2-OH \ \longrightarrow \ CH_3-\overset{\overset{\displaystyle O}{\|}}{C}-O-CH_2CH_2CH_3 \ + \ H_2O$$

acetic acid   1-propanol   propyl acetate

Important Terms

     ester   ester group   esterification   condensation reaction

13.16 Some Important Esters
Besides being used in dynamite, nitroglycerin (an ester of glycerol and nitric acid) has therapeutic value in the treatment of heart disorders. The esters of salicylic acid have many uses. Two examples discussed in this section are methyl salicylate (oil of wintergreen) and acetylsalicylic acid (aspirin).

13.17 Reactions of Esters
Esters are prepared by a condensation reaction between an alcohol and an acid. Esters can be broken apart by the reverse of this reaction, called hydrolysis. In hydrolysis, a molecule of water is added to the ester linkage, breaking it apart to form an alcohol and an acid. Saponification is the breaking of an ester linkage by a strong base. The products of the saponification of an ester are an alcohol and the salt of the acid.

Important Terms

     hydrolysis   saponification

Check Your Understanding ————————————————————————

29.   Name the following esters:

$$(a) \ CH_3CH_2CH_2CH_2CH_2\overset{\overset{\displaystyle O}{\|}}{C}OCH_2CH_3 \qquad (b) \ CH_3CH_2CH_2O\overset{\overset{\displaystyle O}{\|}}{C}H$$

30.   Write the structural formulas for the following compounds:
     (a)   isopropyl benzoate   (b)   octyl heptanoate

For questions 31 to 36, fill in the blank with the correct word or words.
31.   Esters are formed by a _____ reaction between an _____ and an _____.
32.   The reaction in question 32 produces an ester and _____.
33.   Many esters give distinctive flavors to_____.
34.   Methyl salicylate is formed by the condensation reaction between _____ and _____. The common name of methyl salicylate is _____.
35.   Aspirin is formed by the condensation reaction between _____ and _____.
36.   An analgesic is a _____, and an antipyretic is a _____.

For questions 37 to 41, write the equation for the reaction.
37.   The hydrolysis of methyl salicylate.
38.   The condensation reaction between acetic acid and ethanol.
39.   The condensation reaction between benzoic acid and 1-butanol.
40.   The hydrolysis of ethyl butanoate.
41.   The saponification of t-butyl propanoate by potassium hydroxide.

_____

Answers to the Self-Test Questions for Chapter 13

1. (a) $CH_3CH_2OCH_2CH_3$ (b) $CH_3OCH_2CH_2CH_3$ **2.** (a) butyl ethyl ether (b) diisopropyl ether

3. (a) $CH_3CH_2CH_2OH$ (b) $CH_3CHOHCH_3$

(c) $CH_3 - \underset{\underset{CH_3}{|}}{\overset{\overset{CH_3}{|}}{CH}} - \underset{\underset{CH_3}{|}}{\overset{\overset{OH}{|}}{C}} - CH_3$ (d) $CH_3 - CH_2 - \underset{\underset{Br}{|}}{CH} - \underset{\underset{Br}{|}}{CH} - \overset{\overset{OH}{|}}{CH} - CH_3$

4. (a) 2-butanol (b) 2-methyl-2-butanol (c) 1-pentanol (d) 2-methyl-1-propanol
5. (a) secondary (b) tertiary (c) primary (d) primary **6.** ethanol, butanol, hexanol **7.** CH_3OH
8. ethanol, carbon dioxide, energy **9.** dihydric, glycerol **10.** phenol **11.** reduction, hydration
12. propene **13.** primary, secondary
14. (a) $CH_3CH_2CH=CH_2$ (b) $CH_2=CHCH_3$

(c)   ⬡—$CH=CH_2$ (d) $CH_3-\underset{\underset{}{}}{\overset{\overset{CH_3}{|}}{C}}=CH_2$

15. (a) CH_3CH_2CH with $\overset{O}{\overset{\|}{}}$ (b) CH_3CCHCH_3 with $\overset{O}{\overset{\|}{}}$ and CH_3 branch (c) $CH_3CH_2CCH_3$ with $\overset{O}{\overset{\|}{}}$ (d) $CH_3CH_2CHCCH_3$ with $\overset{O}{\overset{\|}{}}$ and CH_3 branch

16. (a) butanal, aldehyde (b) butanone, ketone (c) 3-hexanone, ketone (d) 5-chloropentanal, aldehyde

17. (a) CH_3CH_2OH (b) $CH_3CH_2CH_2COH$ with $\overset{O}{\overset{\|}{}}$ (c) $CH_3CH_2CCH_3$ with $\overset{O}{\overset{\|}{}}$

(d) $CH_3CH_2CHCH_2CH_3$ with OH branch (e) $CH_3CH_2CH_2C\text{-}H$ with OH and OCH_3 branches

(f) $CH_3 - C - CH_3$ with OH and OCH_2CH_3 branches (g) CH_3CHCH_2COH with CH_3 branch and $\overset{O}{\overset{\|}{}}$

18. (a) pentanoic acid (b) 4,5-dimethylhexanoic acid

19. (a) CH_3CH_2COH with $\overset{O}{\overset{\|}{}}$ (b) CH_3CHCH_2COH with $\overset{I}{\overset{|}{}}$ and $\overset{O}{\overset{\|}{}}$ **20.** b **21.** a,d **22.** b **23.** c
24. b **25.** a **26.** d **27.** b **28.** a **29.** (a) ethyl hexanoate (b) propyl formate
30.

(a) benzene ring $-C-O-CH$ with $C=O$, CH_3 and CH_3 branches (b) $CH_3-(CH_2)_5-C-O-CH_2-(CH_2)_6-CH_3$ with $\overset{O}{\overset{\|}{}}$

31. esterification (condensation), carboxylic acid, alcohol **32.** water **33.** fruits **34.** methanol, salicylic acid, oil of wintergreen **35.** salicylic acid, acetic acid **36.** pain reliever, fever reducer

37.

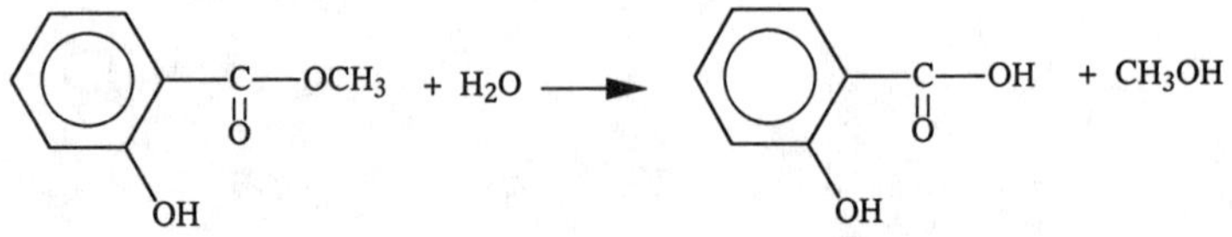

$$\text{(benzene ring)}-\overset{O}{\underset{\text{OH}}{C}}-OCH_3 + H_2O \longrightarrow \text{(benzene ring)}-\overset{O}{\underset{\text{OH}}{C}}-OH + CH_3OH$$

38.

$$CH_3-\overset{O}{C}-OH + CH_3-CH_2-OH \longrightarrow CH_3-\overset{O}{C}-O-CH_2-CH_3 + H_2O$$

39.

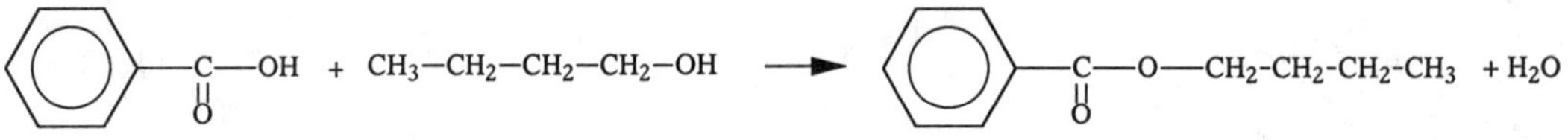

$$\text{(benzene ring)}-\overset{O}{C}-OH + CH_3-CH_2-CH_2-CH_2-OH \longrightarrow \text{(benzene ring)}-\overset{O}{C}-O-CH_2\text{-}CH_2\text{-}CH_2\text{-}CH_3 + H_2O$$

40.

$$CH_3-CH_2-CH_2-\overset{O}{C}-O-CH_2-CH_3 + H_2O \longrightarrow CH_3-CH_2-CH_2-\overset{O}{C}-OH + CH_3-CH_2-OH$$

41.

$$CH_3-CH_2-\overset{O}{C}-O-\underset{\underset{CH_3}{CH_3}}{\overset{CH_3}{C}}-CH_3 + KOH \longrightarrow \underset{OH}{\overset{CH_3}{C}}CH_3-\underset{OH}{C}-CH_3 + CH_3-CH_2-\overset{O}{C}-OK$$

Chapter 14   ORGANIC COMPOUNDS CONTAINING NITROGEN

In this chapter we turn our attention to functional groups that contain nitrogen. Some nitrogen-containing functional groups give basic qualities to the molecule. Nitrogen can also join carbon atoms in ring structures. Nitrogen-containing rings are found in the complex molecules belonging to the class of compounds called alkaloids, which have many physiological effects and are widely used in drug therapy. We discuss a few of these alkaloids at the end of this chapter.

14.1   Functional Groups Containing Nitrogen

Nitrogen is a member of group VA on the periodic table, and is the fourth most abundant element in the human body. In order to achieve an octet nitrogen needs to gain a share of three more electrons. Nitrogen can form three single covalent bonds, or a double and a single covalent bond, or a triple covalent bond to become stable.

14.2   Amines

Amines are compounds containing the amino functional group: $-NH_2$, $-NHR$, or $-NRR'$. They often have strong odors, and are among the products produced by the decay of dead organisms. Amines are classified by the number of carbons bonded to the nitrogen, forming primary, secondary, or tertiary amines.

$$
\begin{array}{ccc}
\text{H---N---H} & \text{H---N---R'} & \text{R''---N---R'} \\
| & | & | \\
\text{R} & \text{R} & \text{R} \\
\text{primary} & \text{secondary} & \text{tertiary}
\end{array}
$$

Amines are polar compounds. Because primary and secondary amines have a hydrogen bonded to nitrogen, they form hydrogen bonds with other amines or water. As a result, amines with short carbon chains are soluble in water.

Amines are named by listing each of the groups attached to the nitrogen and adding the suffix -amine. In more complicated molecules, the $-NH_2$ group is identified by the term -amino. Any groups attached to the amino group are identified by an "N" before the name of the attached group. Thus $CH_3CH_2CH_2NH_2$ is propyl amine. The name of the following more complex amine is N-methylcyclohexane.

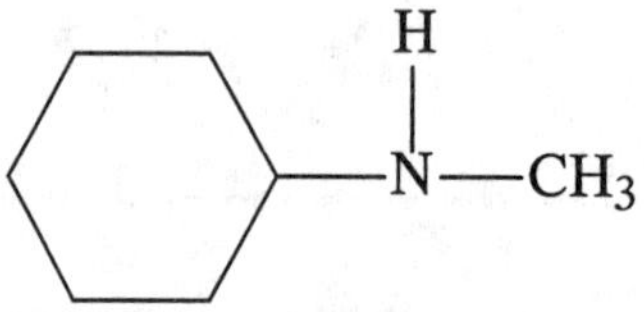

Important Terms

amine amino group

Example

1. Name the following compound:

$$CH_3\text{-}CH_2\text{-}CH_2\text{-}NH\text{-}CH_3$$

S Apply the IUPAC rules for naming to this compound.

T•E The $CH_3CH_2CH_2$- group is the propyl group and the CH_3- group is the methyl group.

P Groups are named alphabetically. Therefore, the name is methylpropylamine.

2. Write the structural formula for 3-(N-methylamino)-butanal.

S Apply the IUPAC nomenclature rules to determine the structural formula of this compound.

T•E Butanal is an aldehyde with four carbon atoms.

```
      O
      ||
C-C-C-C-H
```

The "3-(N-methylamino)" indicates that there is an amino group on carbon 3, and the amino group has a methyl group substituted for one of the hydrogens.

P The structural formula is:

```
   CH3-NH      O
       |       ||
   CH3-CH-CH2-C-H
```

14.3 The Amines as Bases

The unshared pair of electrons on the nitrogen in an amine allows the nitrogen to share electrons with a hydrogen ion. By accepting hydrogen ions, the amine can act as a base. The amines are very similar to ammonia in their basicity. Ammonia is a weak base and the amines are weak bases.

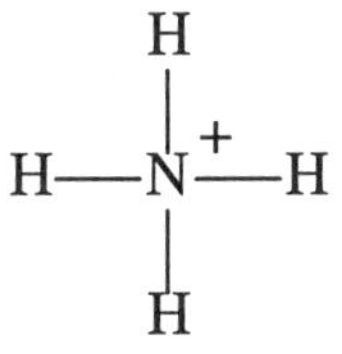

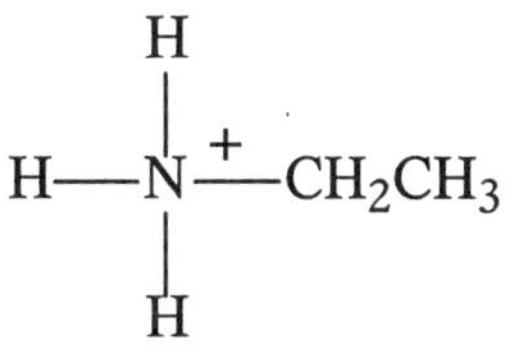

$$H-\overset{\overset{\displaystyle H}{|}}{\underset{\underset{\displaystyle H}{|}}{N}}\overset{+}{}-H \qquad\qquad H-\overset{\overset{\displaystyle H}{|}}{\underset{\underset{\displaystyle H}{|}}{N}}\overset{+}{}-CH_2CH_3$$

The ammonium ion, NH_4^+ is the conjugate acid of ammonia, just as the ethylammonium ion, shown above, is the conjugate acid of ethylamine. The specific group(s) attached to the nitrogen atom will affect the strength of the basic properties of the amine.

14.4 Important Amine Derivatives

Amines will react with acids to form organic salts. When heated with alkyl halides, they form quaternary ammonium salts. Secondary amines react with nitrites to form nitrosamines, compounds that are carcinogenic. Nitrosamines may form in the stomach from nitrites and secondary amines in food. Choline is an important quaternary ammonium salt that is involved in nerve impulse transmission.

Important Term

      quaternary ammonium salt

Check Your Understanding ─────────────────────────────────

1. Draw the structural formula for each of the following compounds:
    (a) hexylamine   (c) ethylmethylpropylamine
    (b) dibutylamine   (d) 2-(N-methylamino)-1-propanol
2. Indicate whether each of the compounds in question 1 is a primary, secondary, or tertiary amine.

3. Name each of the following compounds:

    (a) $CH_3CH_2NHCH_2CH_2CH_3$ (b) $H_2NCH_2CH_2CH_2CH_3$

(c)

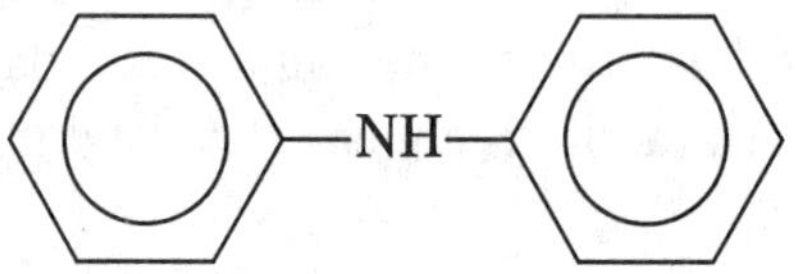

(d)

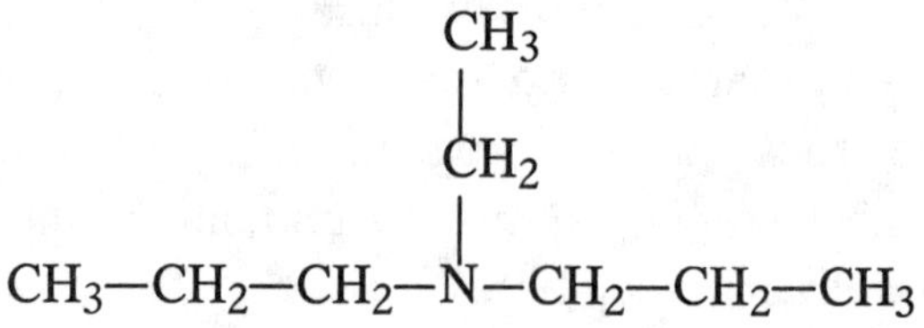

$$CH_3-CH_2-CH_2-\underset{\underset{\underset{CH_3}{|}}{\overset{|}{CH_2}}}{N}-CH_2-CH_2-CH_3$$

4.   Predict the products of the following reactions:

   (a) $CH_3CH_2CH_2NH_2$ + HCl $\longrightarrow$

   (b) $CH_3CH_2NHCH_3$ + H_2O $\longrightarrow$

   (c) $(CH_3CH_2)_3N$ + CH_3CH_2Br $\longrightarrow$

14.5   Amides

Amides are derivatives of carboxylic acids that contain the amide functional group. The bond between the carbon and the nitrogen in the amide group is called an amide linkage and is a very stable bond. Notice the amide linkage in N-methylpropanamide.

$$CH_3-CH_2-\overset{\overset{\displaystyle O}{\|}}{C}-NH-CH_3$$

amide linkage

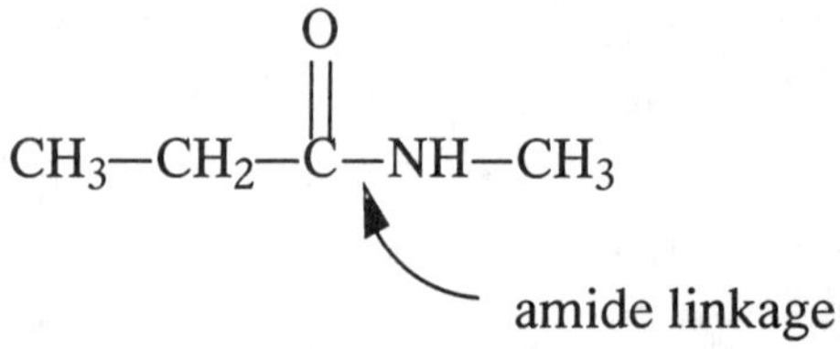

The polar amide group can form hydrogen bonds. As a result, short-carbon-chain amides are soluble in water and have higher melting and boiling points than alkanes with comparable molecular weights.

Simple amides with an unsubstituted nitrogen (nitrogen attached to two hydrogens) are named by changing the ending of the name of the parent acid from "-oic acid" to" -amide". For example, pentanamide.

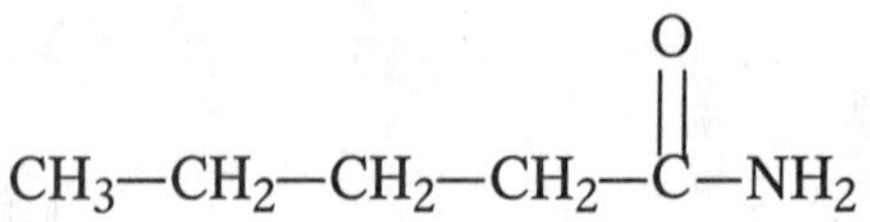

$$CH_3-CH_2-CH_2-CH_2-\overset{\overset{\displaystyle O}{\|}}{C}-NH_2$$

If there are groups attached to the nitrogen, they are identified by placing an "N" before the name of the group. The capital "N" always means nitrogen and indicates that the next group named is bonded to a nitrogen. In N-methylpropanamide, "N-methyl" means that the methyl is bonded to the nitrogen, not to the three carbon chain.

Important Terms

amide   amide linkage

Example ───────────────────────────────────────

1. Name the following amide:

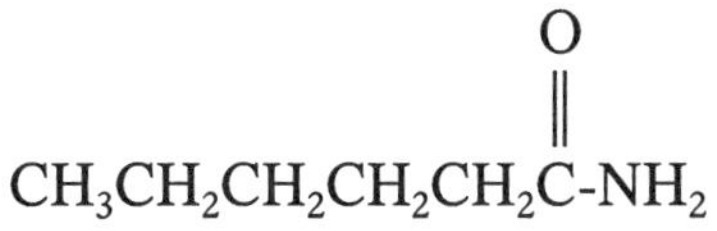

$$O$$
$$\|$$
$$CH_3CH_2CH_2CH_2CH_2C\text{-}NH_2$$

S   Apply the IUPAC rules to name this compound.

T•E This is a simple(unsubstituted) amide whose parent acid had six carbons. The prefix for six carbons is "hex" and the six carbon unbranched acid is hexanoic acid. We drop the "oic acid" and add "amide".

P   Therefore, the name is hexanamide.

2.   Draw the structure of N-ethylbutanamide.

S   Again, apply the IUPAC rules to figure out the structural formula.

T•E The parent acid is butanoic acid which has 4 carbons.

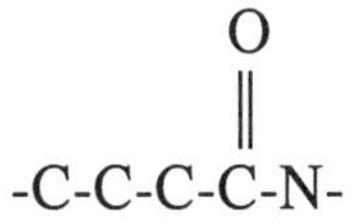

The name tells us that there is one group attached to the nitrogen and it is an ethyl group.

P
$$O$$
$$\|$$
$$CH_3CH_2CH_2C\text{-}N\text{-}CH_2CH_3$$
$$|$$
$$H$$

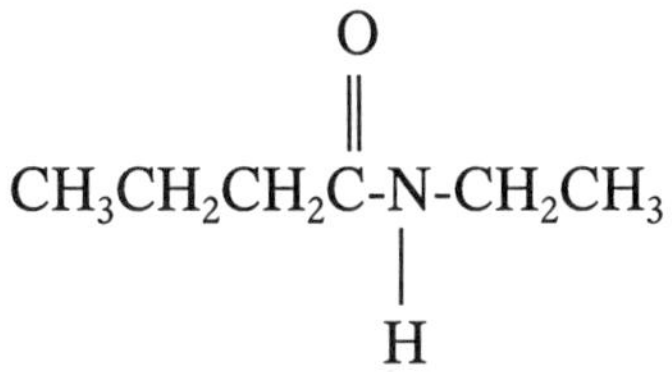

14.6   Some Important Amides

Urea is synthesized from ammonia produced in the breakdown of nitrogen-containing compounds in our bodies. Acetaminophen and the hallucinogen LSD are both amides.

14.7   Reactions of Amides

The amide linkage is very stable, but under certain conditions it can be hydrolyzed or split apart by water. Hydrolysis of an unsubstituted amide produces a carboxylic acid and ammonia. Amides with groups other than hydrogen attached to the nitrogen (N-substituted amides) hydrolyze to form an acid and an amine.

Important Term

   hydrolysis

Example　——

What are the products of the following reaction?

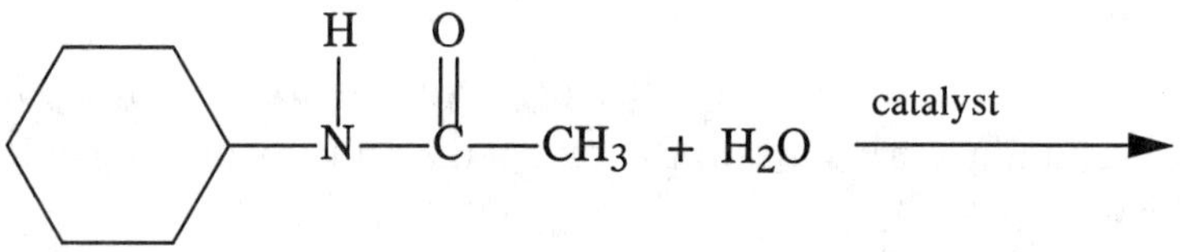

S   The problem asks us to determine what reaction is occurring and to draw the structures of the products.

T•E The organic reactant is an amide. The other reactant is water and the reaction is catalyzed. This is an example of hydrolysis of an amide. Because this is an amide with an alkyl group bonded to the nitrogen it will yield an amine in addition to a carboxylic acid upon hydrolysis.

P   The amide linkage must be broken, a hydrogen attached to the nitrogen and an -OH attached to the carbonyl carbon to give the products.

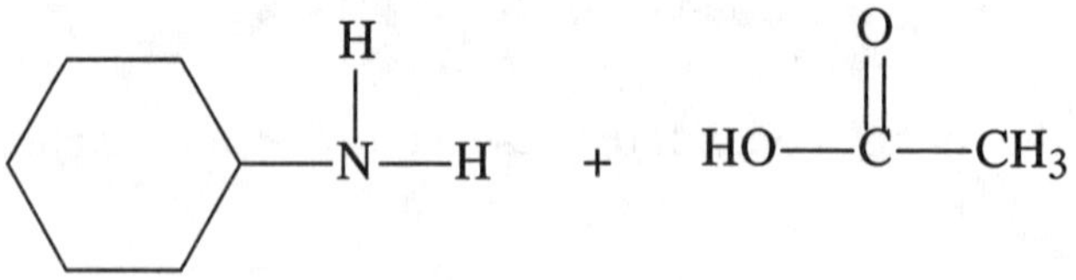

Check Your Understanding　——————————————————————————————————

5.   Which of the following compounds are amides?

(a) $CH_3CCH_2CH_3$

(d) $CH_3CH_2CNHCH_2CH_3$

182

(b) $CH_3CH_2NHCH_2OH$

(e) 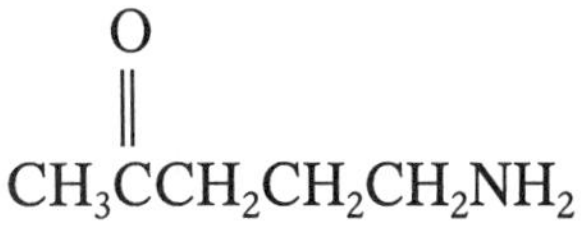 $CH_3\overset{\displaystyle O}{\overset{\|}{C}}CH_2CH_2CH_2NH_2$

(c) $CH_3CH_2\underset{\underset{\displaystyle CH_3}{|}}{CH}\overset{\displaystyle O}{\overset{\|}{C}}NH_2$

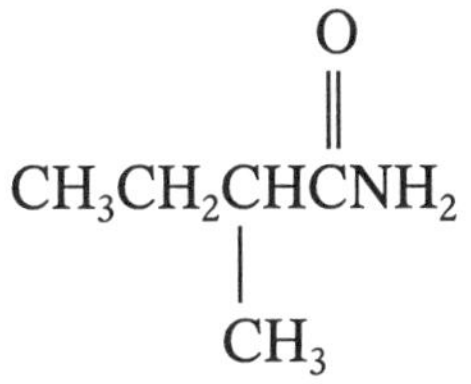

(f) 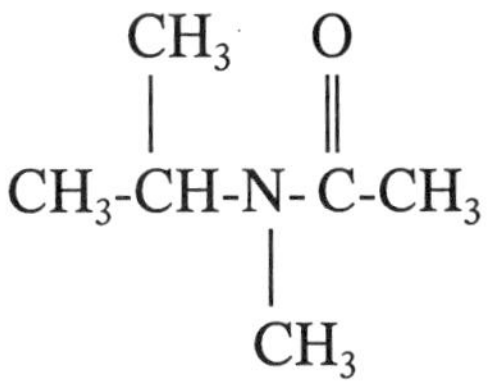 $CH_3-CH-\underset{\underset{\displaystyle CH_3}{|}}{\overset{\overset{\displaystyle CH_3}{|}}{N}}-\overset{\displaystyle O}{\overset{\|}{C}}-CH_3$

6. Name each of the amides in question 5.

7. Write the structural formulas for the following compounds:
   (a)   benzamide
   (b)   N-ethyloctanamide
   (c)   N- ethyl-N-methylhexanamide

8. What are the products of the following reactions?

(a)

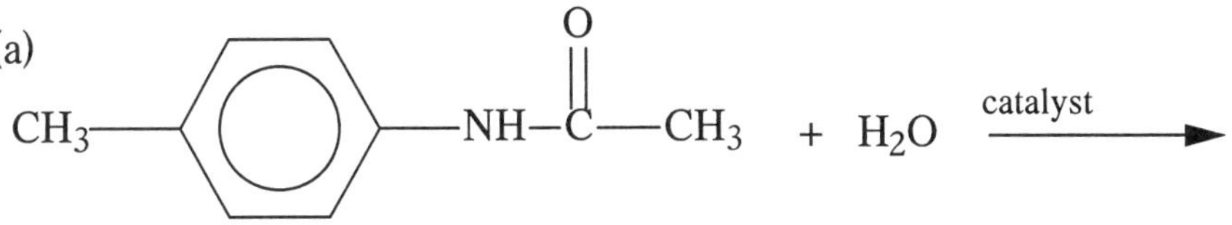

$CH_3-\!\!\bigcirc\!\!-NH-\overset{\displaystyle O}{\overset{\|}{C}}-CH_3$ + H_2O $\xrightarrow{\text{catalyst}}$

(b)

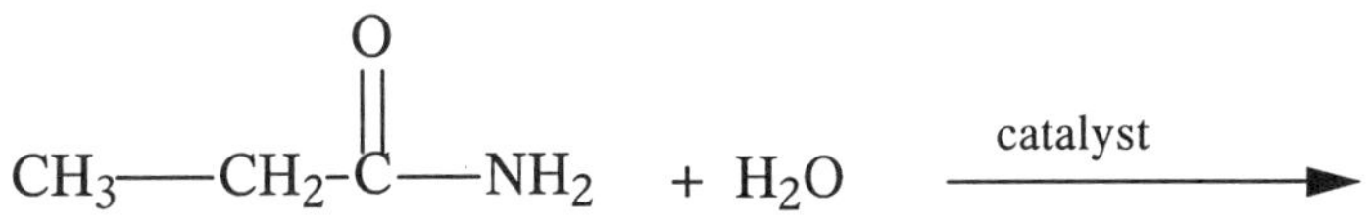

$CH_3-CH_2-\overset{\displaystyle O}{\overset{\|}{C}}-NH_2$ + H_2O $\xrightarrow{\text{catalyst}}$

(c) $CH_3-CH_2-\underset{\underset{\underset{\displaystyle CH_3}{|}}{\overset{\displaystyle CH_2}{|}}}{N}-\overset{\displaystyle O}{\overset{\|}{C}}-CH_2-CH_2-CH_3$ + H_2O $\xrightarrow{\text{catalyst}}$

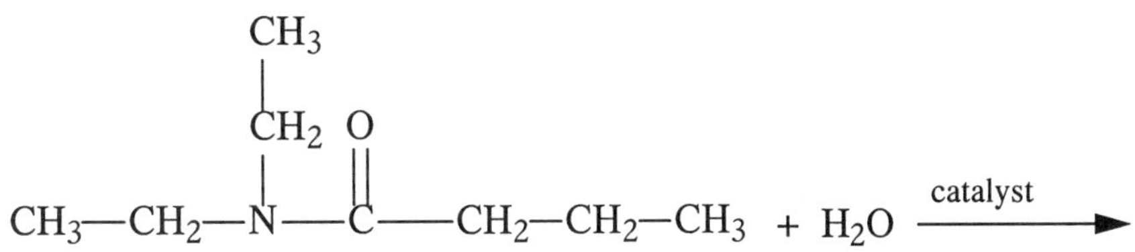

9. Explain with words or a diagram how hydrogen bonding can form between unsubstituted amides such as acetamide, but not between N,N-disubstituted amides such as N,N-dimethylacetamide.

14.8   Heterocyclic Compounds

Rings that contain other elements in addition to carbon and hydrogen are called heterocyclic rings. The number of naturally occurring heterocyclic compounds is very large. We discuss some of these heterocyclic compounds in section 3 of the text. Compare the two cyclic compounds below. The one on the left has a ring made up of only carbon atoms. The one on the right has a nitrogen in its ring and is a heterocycle.

Important Term

> heterocyclic ring

14.9   Alkaloids

The alkaloids are a large class of nitrogen-containing compounds generally forming part of a plant's defense system. Plants containing alkaloids have been used as drugs for many centuries. Many alkaloids have the beneficial properties of reducing pain or producing sedation, but some cause addiction and tolerance on extended use. This section discusses important examples of alkaloids such as epinephrine, norepinephrine, nicotine and the opiates.

Important Terms

> alkaloid   tolerance   addiction

Check Your Understanding ———————————————————————————

For questions 10 to 14, fill in the blank with the correct word or words.

10.   _____ and _____ are two body hormones whose structures differ by only one methyl group.

11.   _____ is an alkaloid found in cigarettes. In pure form it is _____ (mildly, extremely) toxic, and acts on the body by _____ the central nervous system.

12.   _____ and _____ are extracted from opium poppies, and have a _____ effect on humans. _____ is used to reduce pain after a serious accident or surgery. _____ is found in some cough medicines.

13.   _____ is a synthetic alkaloid produced from morphine.

14.   _____ is the name given to the natural substances produced by the brain that produce effects similar to the opiates.

Review Your Understanding ——————————————————————————

Here are twenty multiple choice questions for you to use to review your understanding of the organic chemistry we have covered before you go on to section 3 of the text.

15.   What is the number of hydrogens in the formula for an alkane that has 10 carbons and is not cyclic?
        (a)   10   (c)   12
        (b)   20   (d)   22

16.   Which of the following pairs of compounds are isomers?
        (a) $CH_3CH_2CH_2Cl$ and $CH_3CHClCH_3$
        (b) $CH_3CH_2CH_2CH_2CH_3$ and $CH_3CH_2CH_2CH_3$
        (c) $CH_3(CH_3)CHCH_2CH_2CH_3$ and $CH_3(CH_3)CHCH_2CH_2C(CH_3)_3$
        (d) $CH_3CH_2CCl_2CH_3$ and $CH_2ClCHClCH_2CH_2CH_3$

17.   The complete combustion of ethane produces carbon dioxide plus:
        (a)   CO   (c) H_2O_2
        (b) H_2O (d) $CaCO_3$

18.   Which of these compounds will exhibit cis-trans isomerism?
        (a) $Cl_2C=CCl_2$ (c) $ClHC=CHCl$
        (b) $CH_3CH_2CH_2CH=CH_2$ (d) $CH_2=CHCH_2CH_3$

19.   What is the product of this reaction?

$$CH_3CH_2-C\overset{\displaystyle H}{\underset{\displaystyle}{\|}}=C\overset{\displaystyle H}{\underset{\displaystyle}{|}}-H \;+\; Br_2 \longrightarrow$$

        (a)   2-bromobutane   (c)   1-bromobutane
        (b)   1,2-dibromobutane   (d)   3,4-dibromobutane

20. What is the product of this reaction?

$$CH_3CH_2CH_2CH_2CH=CH_2 \;+\; H_2O \xrightarrow{\;H^+\;}$$

(a) $CH_3-CH_2-CH_2-CH_2-CH=CH-OH$

(c) $CH_3-CH_2-CH_2-\overset{\displaystyle OH}{\underset{\displaystyle |}{C}}H-CH_3$

(b) $CH_3-CH_2-CH_2-CH_2-\overset{\displaystyle OH}{\underset{\displaystyle |}{C}}H-CH_3$

(d) $CH_3-CH_2-CH_2-CH_2-CH_2-CH_3$

21.   What is the product of this reaction?

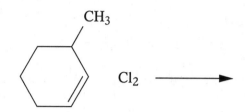

CH$_3$

Cl$_2$ ⟶

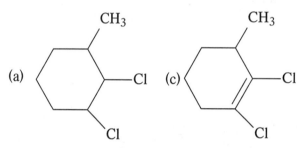

(a)   CH$_3$ ...Cl / Cl   (c)   CH$_3$ ...Cl / Cl

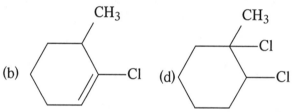

(b)   CH$_3$ ...Cl   (d)   CH$_3$ ...Cl / ...Cl

22.   What is the molecular formula of an eight carbon alkene that has one double bond and is not cyclic?
       (a)   C$_8$H$_8$ (c)   C$_8$H$_{14}$
       (b)   C$_8$H$_{16}$ (d)   C$_8$H$_{18}$

23.   Which of the following must be the molecular formula of a saturated compound?
       (a)   C$_7$H$_{12}$ (c)   C$_6$H$_{12}$
       (b)   C$_7$H$_{14}$ (d)   C$_6$H$_{14}$

24.   A positive Benedict's test indicates the presence of:
       (a)   an alcohol   (c)   an ester
       (c)   an amine   (d)   an aldehyde

25.   Tertiary alcohols are oxidized to form:
       (a)   aldehyes   (c)   ketones
       (b)   carboxylic acids   (d)   none of these

26.   The condensation reaction between a carboxylic acid and an alcohol yields:
       (a)   a ketone   (c)   a ketone plus water
       (b)   an ether plus water   (d)   an ester plus water

27. Aldehydes are synthesized by:
   (a) the reduction of ketones
   (c) the oxidation of secondary alcohols
   (b) the hydration of alkenes
   (d) the oxidation of primary alcohols

28. Identify the functional group in this compound.

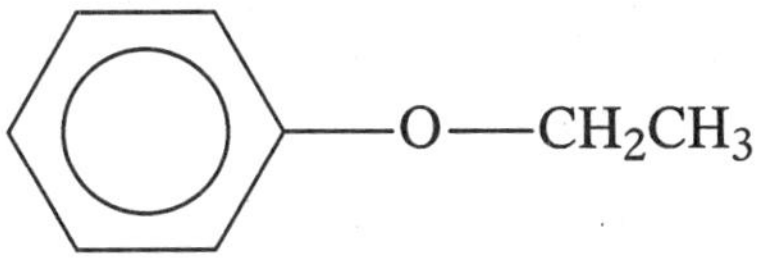

   (a) an ester
   (c) an acetal
   (b) an ether
   (d) an alcohol

29. Identify the functional group in this compound.

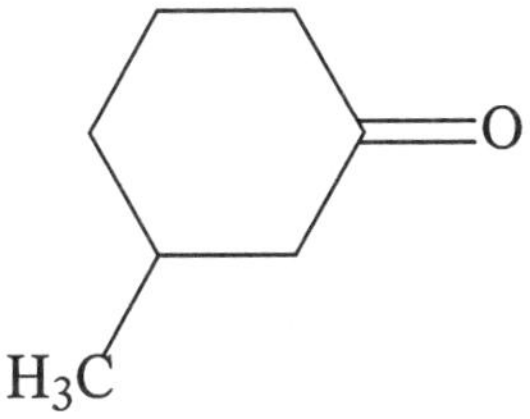

   (a) an ester
   (c) an aldehyde
   (b) an ether
   (d) a ketone

30. The reduction of an aldehyde yields:
   (a) a carboxylic acid
   (c) a ketone
   (b) an acetal
   (d) an alcohol

31. Identify the secondary alcohol

(a) $CH_3-CH_2-CH_2-CH_2-\overset{\displaystyle OH}{\underset{\displaystyle |}{CH}}-CH_3$

(c) $CH_3-CH_2-CH_2-\overset{\displaystyle OH}{\underset{\displaystyle \underset{\displaystyle CH_3}{|}}{\overset{\displaystyle |}{C}}}-CH_3$

(b) $CH_3-CH_2-CH_2-CH_2-CH_2-CH_2-OH$

(d) $CH_3-CH_2-CH_2-\overset{\displaystyle OH}{\underset{\displaystyle \underset{\displaystyle CH_3}{|}}{\overset{\displaystyle |}{C}}}-CH_2-CH_3$

32. The compound $CH_3CH_2NHCH_3$ is:
    (a)   a primary amine   (c)   a secondary amine
    (b)   a tertiary amine   (d)   an amide
33. Identify the isomer of $CH_3CH_2NHCH_3$
    (a) $CH_3CH_2CHOHCH_3$ (c) $CH_3CH_2CH_2NH_2$
    (b) $CH_3CH_2NH_2$ (d) $CH_3CH_2NH_2$
34. Identify the functional group in this compound.

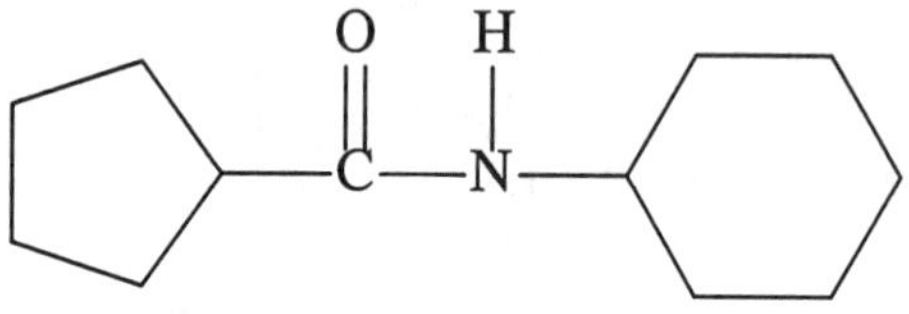

    (a)   an amide   (c)   an amine
    (b)   a ketone   (d)   an ester

Answers to Check Your Understanding Questions in Chapter 14

1. (a) $CH_3(CH_2)_4CH_2NH_2$ (b) $CH_3CH_2CH_2CH_2\text{-}NH\text{-}CH_2CH_2CH_2CH_3$

(c) $CH_3CH_2\text{-}N\text{-}CH_2CH_2CH_3$ (d) $CH_3\text{-}CH\text{-}CH_2OH$
        | |
      CH_3 $CH_3\text{-}N\text{-}H$

2. (a) primary (b) secondary (c) tertiary (d) secondary

3. (a) ethylpropylamine (b) butylamine (c) diphenylamine (d) ethyldipropylamine

4. (a) $CH_3CH_2CH_2NH_3{}^+Cl^-$ (b) $CH_3CH_2NH_2{}^+OH^-$ (c) $(CH_3CH_2)_4N^+Br^-$
                                    |
                                 CH_3

5. c,d,f   **6.** (c) 2-methylbutanamide (d) N-ethylpropanamide
(f) N-isopropyl-N-methylacetamide

7. (a)

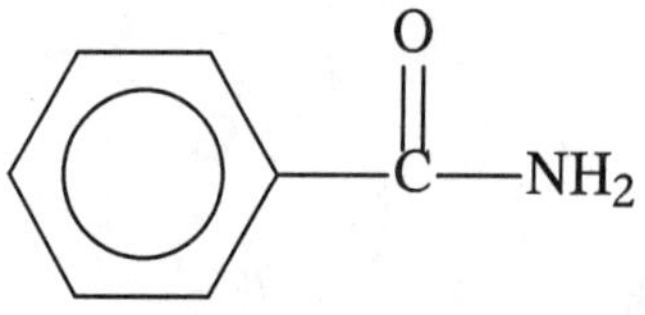

(b) $CH_3-(CH_2)_6-\overset{\displaystyle O}{\overset{\|}{C}}-NH-CH_2-CH_3$

(c) $CH_3-CH_2-CH_2-CH_2-CH_2-\overset{\displaystyle O}{\overset{\|}{C}}-\overset{\displaystyle CH_3}{\overset{|}{N}}-CH_2CH_3$

8.

(a) $CH_3-\underset{}{\bigcirc}-NH_2 + CH_3-\overset{\displaystyle O}{\overset{\|}{C}}-OH$

(b) $CH_3-CH_2-\overset{\displaystyle O}{\overset{\|}{C}}-OH + NH_3$

(c) $CH_3-CH_2-NH-CH_2-CH_3 + CH_3-(CH_2)_2-\overset{\displaystyle O}{\overset{\|}{C}}-OH$

9. The nitrogen on acetamide has two hydrogen atoms attached to it that can form hydrogen bonds with oxygens on other acetamide molecules. N,N-dimethylacetamide has no hydrogen atoms attached to the nitrogen and, therefore, cannot form hydrogen bonds.

10. epinephrine, norepinephrine **11.** nicotine, extremely, stimulating **12.** morphine, codeine, narcotic, morphine, codeine **13.** heroin **14.** endorphins

15. d **16.** a **17.** b **18.** c **19.** b **20.** b **21.** a **22.** b **23.** d **24.** d **25.** d **26.** d **27.** d **28.** b **29.** d **30.** d **31.** a **32.** c **33.** c **34.** a

Chapter 15   CARBOHYDRATES

In this chapter we begin studying the compounds of life by discussing the large class of compounds called carbohydrates. Compounds in this class include sugars, starches, and cellulose. Carbohydrates are found mainly in plant material; they function both as the supporting structure and food storage molecules of the plant. Carbohydrates can be classified by the size of the molecule.

15.1   Classification

Monosaccharides are simple sugars that cannot be broken into smaller units upon hydrolysis. Disaccharides contain two monosaccharides; they will break apart to form two simple sugars upon hydrolysis. Polysaccharides are polymers containing many monosaccharide units.

Monosaccharides may be further classified by the number of carbon atoms in the molecule and by the carbonyl functional group found in the molecule. Hexoses contain six carbon atoms, while pentoses contain five. If the carbonyl group in the monosaccharide is an aldehyde, the sugar is an aldose. Ketoses contain the ketone group. An aldopentose is a five carbon aldehyde-containing monosaccharide.

Important Terms

monosaccharide	disaccharide	polysaccharide
aldose	ketose	

Check Your Understanding _____

For questions 1 to 7, fill in the blanks with the correct word or words.
1.   _____ is the name given to the large class of compounds that includes sugars, starches, and gums.
2.   These compounds are found mainly in _____, and their functions are _____ and _____.
3.   _____ yield three or more simple sugars upon hydrolysis.
4.   _____ are carbohydrates that cannot be broken down by hydrolysis.
5.   _____ yield two simple sugars upon hydrolysis.
6.   A monosaccharide that contains five carbon atoms and a ketone functional group is a _____.
7.   A monosaccharide that contains six carbon atoms and an aldehyde functional group is an _____.

15.2 Monosaccharides: Glucose

Simple sugars are white crystalline solids that are soluble in water and have a sweet taste. The most abundant monosaccharides are the hexoses, and the most abundant hexose is glucose. Glucose is an aldohexose. It is the immediate energy source for cells in living organisms and is the monomer unit for many important polysaccharides.

Glucose is found in three forms in aqueous (water) solution. Two different cyclic (ring) forms (alpha and beta) and a straight chain form exist in equilibrium with one another, with the cyclic forms being much more common. We can write the structure of glucose in straight chain form, or in ring form using Haworth projections. The ring results from the formation of an internal hemiacetal by the reaction between the aldehyde group on carbon number 1 and the hydroxyl group on carbon number 5. The formation of a hemiketal between the ketone group and a hydroxyl group results in similar ring forms in the ketoses.

Study the structures of glucose in Section 15.2 in the text. Notice those groups to the right on carbons 2, 3 and 4 in the open chain form are placed below the plane of the ring in the cyclic form. The -OH on carbon 1 is above the plane of the ring in the β form and below the plane of the ring in the α form. Be certain that you can match the carbons in the cyclic form with the corresponding carbons in the open chain form. The -CH₂OH group, for example, is carbon 6.

Important Terms

hexose	aldohexose	glucose
hemiacetal	hemiketal	dextrose
Haworth projection	alpha (α) ring form	beta (β) ring form

Check Your Understanding —————————————————————————————

For questions 8 to 10, fill in the blanks with the correct word or words.
8. The monosaccharide _____ is also called dextrose.
9. The major function of glucose in living cells is _____.
10. The three structural forms of glucose in water solution are _____, _____, and
    _____.

11. Which of these structures is:
    (a) the alpha form of glucose
    (b) the beta form of glucose

191

15.3 Optical Isomerism

A carbon atom that has four different groups attached to it is called a chiral carbon, and will possess a property called chirality. The four groups might be as obviously different as chlorine and hydrogen or they might be only slightly different, such as isopropyl and propyl.

A molecule that cannot be superimposed on its mirror image exhibits chirality. Chirality means "handedness." The molecule and its nonsuperimposable mirror image are called enantiomers. These mirror image molecules are optical isomers that will rotate the plane of polarized light in opposite directions. Enantiomers have identical physical properties except one: they will interact differently with polarized light. Their chemical properties will be identical except when they react with other chiral molecules. The relationship between enantiomers is analogous to the similarities and dissimilarities between your right and left hands or between your feet. Try placing a left shoe on your right foot and you will experience one example of chirality. Your foot and the shoe are both chiral, so you must put the right shoe on the right foot. Socks, however, are usually not chiral. You can put either one of most pairs of socks on either foot.

Carbohydrate molecules are optically active. The D and L classification system of optical isomers is based on the three-carbon compound glyceraldehyde. Most naturally occurring carbohydrates belong to the D family of isomers. In the D family, the hydroxyl group on the last chiral carbon is placed to the right of the chain. In the L family, it is placed to the left.

Important Terms

      optical isomerism   optical isomer   chirality
      chiral center   chiral molecule   enantiomers

Check Your Understanding _____

For questions 12 to 19, answer true (T) or false (F).

12.   When bonded to four different groups, a carbon atom possesses a property called chirality.

13.   Carbon 2 in 2-butanol is a chiral center.

$$CH_3 - \overset{\displaystyle OH}{\underset{\displaystyle H}{\overset{|}{\underset{|}{C}}}} - CH_2CH_3 \quad \text{(2-butanol)}$$

14.   A chiral molecule can be superimposed on its mirror image.

15. Polarized light is light that vibrates in several planes.
16. Molecules that are nonsuperimposable mirror images of each other have the same physical and chemical properties.
17. Enantiomers both rotate the plane of polarized light in a clockwise direction.
18. If a chiral molecule rotates the plane of polarized light in a clockwise direction, its mirror image will rotate the plane of polarized light in a counterclockwise direction.

15.4 Fructose

Fructose is a ketohexose that, like glucose, exists in alpha and beta ring forms. Carefully study the open chain and cyclic forms of fructose in the text. Pay special attention to the numbering of the carbons in the cyclic form. You can tell the difference between carbon 2 and carbon 5 in the cyclic form because carbon 2 has a hydroxyl group bonded to it, while carbon 5 does not. We know that the isomer shown is the α isomer because the hydroxyl on carbon 2 is below the plane of the ring. When combined with glucose, fructose forms the disaccharide sucrose. Fructose is found in fruit juices and honey.

Important Terms

fructose   ketohexose   hemiketal

15.5 Galactose

Galactose is an aldohexose that is not found in nature as a free monosaccharide. It is a component of larger molecules such as lactose, the glycolipids, and agar-agar. Look closely at the structures of glucose and galactose in Table 15.1. The only difference is at carbon 4. The hydroxyl group on this carbon is to the left of the chain in galactose and to the right of the chain in glucose.

Important Terms

galactose   lactose

15.6 Pentoses

Pentoses are five-carbon sugars. Important pentose sugars are arabinose, xylose, ribose, and deoxyribose.

Important Terms

pentose   arabinose   xylose
ribose   deoxyribose

15.7 Reducing Sugars

Carbohydrates that will reduce basic solutions of mild oxidizing agents such as Cu^{2+} or Ag^+ are called reducing sugars. In order to be a reducing sugar the carbohydrate must have a free, or potentially free, aldehyde or ketone group. All monosaccharides are either in the open chain form or they are in the hemiacetal or hemiketal form that is in equilibrium with the open chain form. Therefore, all monosaccharides are reducing sugars.

Benedict's test is very useful for detecting the presence of a reducing sugar. A reducing sugar will reduce the Cu^{2+} ions in the Benedict's test solution to Cu^+ ions, resulting in the formation of a brick-red precipitate of Cu_2O. Clinitest tablets make use of this reaction to test for sugar in urine. Tollens's pentose test can be used to decide if the sugar found using the Benedict's test is a hexose or pentose.

Important Terms

reducing sugar   Benedict's test   Tollens's pentose test

Check Your Understanding ────────────────────────────────────

For questions 19 to 35, select the best answer or answers from the choices given.

19.   Which of the following is an aldohexose?
   (a)   ribose   (c)   fructose
   (b)   glucose   (d)   arabinose

20.   Carbohydrates that contain two simple sugars are called
   (a)   monosaccharides   (c)   disaccharides
   (b)   polysaccharides   (d)   glycolipids

21.   The sugar sucrose contains the simple sugars
   (a)   ribose   (c)   fructose
   (b)   glucose   (d)   galactose

22.   One of the simple sugars found in lactose is
   (a)   ribose   (c)   fructose
   (b)   galactose   (d)   xylose

23.   Tests used to identify the presence of a reducing sugar:
   (a)   Tollens's solution   (c)   Benedict's solution
   (b)   Clinitest   (d)   Fehling's solution

24.   A test used to distinguish between a hexose and pentose sugar:
   (a)   Tollens's pentose   (c)   Benedict's solution
   (b)   Clinitest   (d)   Fehling's solution

25.   Which of the following is a ketohexose?
   (a)   ribose   (c)   fructose

(b)   glucose   (d)   galactose

26.   The name given to the ring structure of glucose when the hydroxyl group on carbon 1 is above the ring is

      (a)   alpha form   (c)   beta form

      (c)   hemiacetal   (d)   hemiketal

27.   A polymer of which monosaccharide is found in agar-agar?

      (a)   ribose   (c)   fructose

      (b)   glucose   (d)   galactose

28.   A carbohydrate that has a free aldehyde or ketone group is called a

      (a)   reducing sugar   (c)   disaccharide

      (b)   polysaccharide   (d)   hemiacetal

29.   Which of the following is an aldopentose?

      (a)   ribulose   (c)   ribose

      (b)   galactose   (d)   fructose

30.   The name given to the ring structure of galactose when the hydroxyl group on carbon 1 is below the ring is

      (a)   alpha form   (c)   beta form

      (c)   hemiacetal   (d)   hemiketal

31.   The five-member ring structure of fructose is called a(n)

      (a)   acetal linkage   (c)   ketohexose

      (b)   hemiacetal   (d)   hemiketal

32.   Another name for blood sugar is

      (a)   glucose   (c)   ribose

      (b)   galactose   (d)   fructose

33.   Carbohydrates that contain many simple sugars are called

      (a)   monosaccharides   (c)   disaccharides

      (b)   polysaccharides   (d)   glycolipids

34.   The six-member ring structure of glucose is called a(n)

      (a)   acetal linkage   (c)   aldohexose

      (b)   hemiacetal   (d)   hemiketal

35.   A carbohydrate that cannot be broken into smaller units upon hydrolysis is called a

      (a)   monosaccharide   (c)   disaccharide

      (b)   polysaccharide   (d)   glycolipid

15.8   Disaccharides: Maltose

As seen earlier, a hemiacetal can react with a molecule of an alcohol to form an acetal. By the same process, hemiketals can form ketals. The formation of disaccharides involves the conversion of hemiacetals or hemiketals into acetals or ketals.

To describe the structure of a disaccharide you must know three things: the two monosaccharide molecules it yields on hydrolysis, the positions on the two monosaccharides that are linked and whether the linkage is α or β. Maltose (or malt sugar) is a disaccharide containing two units of glucose connected by an α,1:4 linkage. This bond is a glycosidic (or acetal) linkage formed between carbon 1 on one glucose molecule in the alpha form and carbon 4 on the second glucose molecule. The aldehyde group on the second glucose molecule is not involved in the acetal linkage; therefore, maltose is a reducing sugar.

Important Terms

maltose	acetal linkage	condensation reaction
glycosidic linkage	α,1:4 linkage	

15.9 Lactose
Lactose is a disaccharide containing a glucose unit and a galactose unit. The acetal linkage is between carbon 1 of galactose in the beta form and carbon 4 of glucose, and is called a β,1:4 linkage. Lactose is found in the milk of mammals. The aldehyde group on the glucose is not involved in the acetal linkage; therefore, lactose is a reducing sugar.

Important Term

lactose	β,1:4 linkage

15.10 Sucrose
Sucrose (or table sugar) is a disaccharide containing a unit of glucose and a unit of fructose. It is the sugar we use in cooking. The linkage in the sucrose molecule is an α,1:2 linkage between the aldehyde group on the glucose and the ketone group on the fructose. The glucose has become an acetal and the fructose has become a ketal. Therefore, sucrose is not a reducing sugar.

Important Terms

sucrose	α,1:2 linkage

Check Your Understanding _____

For questions 36 to 41, fill in the blank with the correct word or words.
36. An _____ linkage is formed in a condensation reaction between a hemiacetal and an alcohol.
37. This linkage is _____ (more, less) stable than a hemiacetal linkage.
38. _____ is composed of glucose and fructose.

39. Maltose contains two _____ units held together by a _____ linkage.
40. _____ contains a unit of galactose and a unit of glucose, and is found in _____.
41. When produced in the controlled germination of grains, _____ (also called malt) is used in the production of _____.

15.11 Polysaccharides: Starch

Starch is a polymer of glucose used by plants as the storage form for energy. Starch is a mixture of two polysaccharides: amylose and amylopectin. Amylose is a linear polysaccharide in which all the glucose units are connected by $\alpha,1{:}4$ linkages. Amylopectin is a branched polysaccharide. Besides glucose units connected by $\alpha,1{:}4$ linkages, as found in amylose, it also has $\alpha,1{:}6$ linkages at the branch points.

Dextrins are polysaccharides formed by the partial hydrolysis of starch. They are formed in the baking of bread, and are used as adhesives on stamps and envelopes.

Important Terms

    starch   amylose   amylopectin
    $\alpha,1{:}6$ linkage   dextrins

15.12 Glycogen

Glycogen is a highly branched polymer of glucose that is the storage form of glucose in animals. As in amylopectin, the branching is a result of $\alpha,1{:}6$ linkages between the glucose units. The level of glucose in the blood remains relatively constant because the body stores excess glucose as glycogen in the liver right after a meal and then slowly releases the glucose back into the blood as the level falls.

Important Term

    glycogen

15.13 Cellulose

Cellulose is a polymer of glucose produced by plants to form the supportive framework of the plant. The rigidity that cellulose gives to plants is a result of hydrogen bonding between cellulose molecules. The glucose units in cellulose are held together by $\beta,1{:}4$ linkages. We do not have enzymes in our digestive tract that can break apart $\beta,1{:}4$ linkages between glucose molecules. Therefore, any cellulose that we eat passes through our digestive tract undigested.

Important Terms

    cellulose   β,1:4 linkage

15.14 Iodine Test

The iodine test is used to detect the presence of starch. It will turn an intense blue-black color when starch is present.

Important Term

    iodine test

Check Your Understanding ————————————————————————

For questions 42 to 58, match the statement with the correct answer or answers from the list. An answer may be used more than once.

A. acetal	D. sucrose	G. amylose	J. cellulose
B. maltose	E. starch	H. amylopectin	K. iodine test
C. lactose	F. glycogen	I. dextrin	

42.   Hemiacetal and an alcohol
43.   Contains galactose and glucose
44.   Storage form of glucose in plants
45.   Contains α,1:6 linkages
46.   Forms support structures in plants
47.   Identifies the presence of starch
48.   Used in the manufacture of beer
49.   Found in the milk of mammals
50.   Contains glucose and fructose
51.   Storage form of glucose in animals
52.   Forms golden brown color of bread crusts
53.   Contains β,1:4 linkages between glucose molecules
54.   Contains two glucose units
55.   Table sugar
56.   Polysaccharides forming starch
57.   Used as an adhesive
58.   Used to produce plastics, films, and guncotton

15.15 Photosynthesis

Photosynthesis is the process by which plants capture the sun's energy and use it to produce carbohydrates. The process is not entirely understood, but it involves two series of reactions: the light and dark reactions. The light reactions require sunlight and molecules of chlorophyll, and produce oxygen and energy-rich molecules. The dark reactions then use the energy-rich molecules to form glucose and other organic molecules from carbon dioxide. The overall reaction for photosynthesis is:

$$6CO_2 + 6H_2O \xrightarrow[\text{chlorophyll}]{\text{light}} C_6H_{12}O_6 + 6O_2$$

This reaction is part of the carbon cycle, by which carbon is continuously being exchanged among the earth, the atmosphere and living organisms. In the reverse of this reaction, cells convert glucose to carbon dioxide and water and produce the energy necessary to continue the life processes.

Important Terms

photosynthesis   chloroplast   chlorophyll   carbon cycle

Check Your Understanding ─────────────────────────────

For questions 59 to 63, fill in the blanks with the correct word or words.
59.   _____ is the process by which plants produce carbohydrates.
60.   This process involves two series of reactions called _____ and _____ reactions.
61.   The _____ reactions require the presence of chlorophyll and sunlight, and produce _____ and _____.
62.   The _____ reactions, which can occur without sunlight, produce _____ from _____ and _____.
63.   All these reactions take place in the _____ of the plant cell.

Answers to Check Your Understanding Questions in Chapter 15

1. carbohydrate **2.** plants, support, energy storage **3.** polysaccharides **4.** monosaccharides
5. disaccharides **6.** ketopentose **7.** aldohexose **8.** glucose **9.** supply immediate energy needs
10. alpha ring, beta ring, straight chain **11.** (a) II (b) I **12.** T **13.** T **14.** F **15.** F **16.** F **17.** F
18. T **19.** b **20.** c **21.** b, c **22.** b **23.** b, c, d **24.** a **25.** c **26.** c **27.** d **28.** a **29.** c **30.** a **31.** d
32. a **33.** b **34.** b **35.** a **36.** acetal or glycosidic **37.** more **38.** sucrose **39.** glucose, α,1:4
40. lactose, milk of mammals **41.** maltose, beer **42.** A **43.** C **44.** E **45.** F, H **46.** J **47.** K
48. B **49.** C **50.** D **51.** F **52.** I **53.** J **54.** B **55.** D **56.** G, H **59.** I **60.** J **61.** photosynthesis
60. light, dark **61.** light, oxygen, energy-rich molecules **62.** dark, glucose, carbon dioxide,
energy-rich molecules **63.** chloroplasts

Chapter 16   LIPIDS

We now turn to a second major class of the compounds of life: the lipids. In the last chapter, we saw that carbohydrates are highly polar substances, and that monosaccharides and disaccharides are soluble in water. Lipids are oily or waxy substances that function as energy storage molecules and as structural components of cells. The lipids include all biological compounds that are not soluble in water but are soluble in organic solvents such as carbon tetrachloride. They are found in the structural components of cells. A second major function of lipids is the storage of energy in cells.

16.1   Saponifiable Lipids: Simple Lipids

Lipids can be categorized as saponifiable (those that can be hydrolyzed by a base) and nonsaponifiable. The saponifiable lipids are the simple lipids that yield fatty acids and alcohol upon hydrolysis. The compound lipids, when hydrolyzed, yield fatty acids, alcohol, and other compounds.

$$\text{simple lipids + base} \longrightarrow \text{fatty acids + alcohol}$$
$$\text{compound lipids + base} \longrightarrow \text{fatty acids + alcohol + other compounds}$$

Important Terms

simple lipid   compound lipid   saponifiable

16.2   Fats and Oils

Glycerol (1,2,3-propanetriol) has three hydroxyl groups and can react with three acids to form a triester called a triacylglycerol. These compounds are also called neutral fats or triglycerides. In simple triacylglycerols, all three fatty acids are the same. Mixed triacylglycerols contain two or more different fatty acids. Natural fats are a mixture of simple and mixed triacylglycerols. They function as energy storage molecules, and include both simple and mixed triacylglycerols.

Important Terms

neutral fat   simple triacylglycerol   triglyceride
triacylglycerol   mixed triacylglycerol

16.3   Fatty Acids

Fatty acids are long-carbon-chain carboxylic acids that are formed upon hydrolysis of triacylglycerols. Notice the melting points of the fatty acids in Table 16.2 in the text.

Saturated fatty acids contain only carbon-to-carbon single bonds and are solids at room temperature. Unsaturated fatty acids have one or more carbon-to-carbon double bonds and are liquids at room temperature. Fats and oils differ in the number of unsaturated fatty acids they contain. We can show the position and number of double bonds in an unsaturated fatty acid by a shorthand code, as shown at the bottom of Table 16.2.

Example _____

What is the shorthand code for the following acid?

$$CH_3CH_2CH_2CH_2CH_2CH=CHCH_2CH=CHCH_2CH_2CH_2CH_2CH_2CH_2CH_2COOH$$

S   Apply the shorthand code rules to this fatty acid.

T   Count the number of carbons, then count the number of carbon-to-carbon double bonds and then count the number of carbons from the methyl end of the structure to the first double bond.

E   The molecule contains 18 carbons, there are two double bonds and the first double bond is 6 carbons from the methyl end of the molecule.

P   The shorthand code for this acid is: 18:2n-6.
       Can you identify this fatty acid using Table 16.2? It is linoleic acid.

Important Terms

unsaturated fatty acid	fatty acid	oil
saturated fatty acid	fat	

16.4   Essential Fatty Acids

Our bodies cannot synthesize adequate amounts of certain fatty acids known as essential fatty acids. These fatty acids are required for the health and growth of tissues. As a result, it is important that our diets contain an adequate supply of linoleic and arachidonic acids. The body uses arachidonic acid as the starting material to synthesize a very important class of compounds called eicosanoids.

Important Terms

essential fatty acid	eicosanoids

Check Your Understanding _____

For questions 1 to 16, choose the correct answer or answers.

1. Another name for triacylglycerol is
   (a) triglyceride
   (c) simple lipid
   (b) neutral fat
   (d) compound lipid
2. A fatty acid that contains only carbon-to-carbon single bonds is
   (a) saponifiable
   (c) nonsaponifiable
   (b) saturated
   (d) unsaturated
3. The essential fatty acids are
   (a) palmitic
   (c) arachidonic
   (b) linoleic
   (d) stearic
4. The hydrolysis of a triacylglycerol produces
   (a) fatty acids
   (c) simple lipids
   (b) compound lipids
   (d) glycerol
5. Contains many unsaturated fatty acids.
   (a) compound lipid
   (c) oil
   (b) fat
   (d) simple triacylglycerol
6. A lipid that can be hydrolyzed by a base is
   (a) saponifiable
   (c) nonsaponifiable
   (b) saturated
   (d) unsaturated
7. These compounds yield fatty acids, alcohol, and other compounds upon hydrolysis.
   (a) simple lipids
   (c) simple triacylglycerols
   (b) compound lipids
   (d) mixed triacylglycerols
8. A neutral fat that contains different fatty acids is called a
   (a) simple lipid
   (c) simple triacylglycerol
   (b) compound lipid
   (d) mixed triacylglycerol
9. One of the most abundant fatty acids is
   (a) butyric acid
   (c) arachidic acid
   (b) linolenic acid
   (d) oleic acid
10. The tissue that contains fat cells is called
    (a) adipose tissue
    (c) epithelial tissue
    (b) connective tissue
    (d) muscle tissue
11. The body requires these in the diet
    (a) triacylglycerols
    (c) essential fatty acids
    (b) compound lipids
    (d) saturated fatty acids
12. Compounds that contain only fatty acids and alcohol.
    (a) simple lipids
    (c) simple triacylglycerols
    (b) compound lipids
    (d) mixed triacylglycerols
13. A fat that has three identical fatty acids is called a
    (a) simple lipid
    (c) simple triacylglycerol
    (b) compound lipid
    (d) mixed triacylglycerol

14. A fatty acid that contains carbon-to-carbon double bonds is
    (a) saponifiable   (c) nonsaponifiable
    (b) saturated   (d) unsaturated
15. This compound is used by the body to synthesize prostaglandins.
    (a) palmitic acid   (c) arachidonic acid
    (b) oleic acid   (d) stearic acid
16. This lipid contains more saturated than unsaturated fatty acids.
    (a) compound lipid   (c) oil
    (b) fat   (d) simple triacylglycerol

16.5   Waxes
Waxes are esters of long-chain fatty acids and long-chain alcohols. They are insoluble in water, flexible, and nonreactive. They form various types of protective coatings on plants and animals.

Important Term

wax

16.6   Iodine Number
The iodine number measures the degree of unsaturation of a triacylglycerol. The higher the iodine number, the greater the unsaturation. Fats have iodine numbers below 70, and oils above 70. Table 16.3 lists the iodine number for several common fats and oils.

Important Term

iodine number

16.7   Hydrogenation
Hydrogenation is the process of adding hydrogen to the double bonds in the fatty acids of an oil, converting it to a solid fat. This process is used commercially to produce solid shortenings and margarine. There is considerable concern about the consumption of the trans fatty acids produced by the hydrogenation of oils.

16.8   Rancidity
A fat becomes rancid when it undergoes hydrolysis or oxidation. Butter fat can be hydrolyzed by microorganisms in the air, releasing the strong-smelling butyric acid. Oxygen in the air can oxidize unsaturated fats and oils to form short-chain fatty acids and aldehydes having disagreeable odors. This oxidation can be slowed by adding antioxidants to the food product.

Hydrolysis

Fat/Oil + H_2O $\longrightarrow$ fatty acids + glycerol

Oxidation

Fat/Oil + O_2 $\longrightarrow$ acids + aldehydes

Important Terms

rancidity   antioxidant

16.9 Hydrolysis

Hydrolysis of triacylglycerols occurs in the presence of steam, hot mineral acids, or enzymes. This hydrolysis yields three fatty acids and glycerol.

Important Term

hydrolysis

16.10 Saponification

Hydrolysis of triacylglycerols in the presence of a strong base yields glycerol and three fatty acid salts. Such salts are soaps. The type of metal ion in the salt and the amount of unsaturation in the fatty acid will determine the nature of the soap. Calcium, magnesium, and iron soaps are insoluble in water. These ions are often found in hard water and will cause the soap to precipitate, forming soap scum and decreasing the cleansing action of the soap. The calcium and magnesium salts of synthetic detergents are soluble in water, and they are not affected by pH. This makes synthetic detergents more effective than soaps in hard or acidic water.

A soap molecule contains a nonpolar tail that can dissolve in grease and oil, and a polar head that is soluble in water. Soap acts by emulsifying grease, making it form small colloidal particles that can be washed away by water.

Important Terms

saponification   soap   detergent
hard water   hydrophobic   hydrophilic

Check Your Understanding ———————————————————————————

For questions 17 to 24, fill in the blanks with the correct word or words.

17.   The iodine test measures the _____ of a triacylglycerol.

18.   _____ have iodine numbers greater than 70, and _____ have iodine numbers less than 70.

19.   _____ is the process by which hydrogen is added to the double bonds in a triacylglycerol, forming a _____ fat from a _____ fat.

20.   If arachidonic acid were treated by the process in question 19, _____ acid would be produced.

21.   When a fat or oil undergoes oxidation, it may develop a disagreeable _____ or _____, and is called _____.

22.   Hydrolysis of a simple triacylglycerol yields _____ and _____.

23.   _____ is the hydrolysis of a triacylglycerol in the presence of a strong base; it produces _____ and _____, which are called soaps.

24.   Liquid soaps are produced when the hydrolysis is carried out in the presence of _____; solid soaps result when the hydrolysis is carried out in the presence of _____.

For questions 25 to 32, answer true (T) or false (F).

25.   Soap can emulsify grease.

26.   Magnesium soaps are soluble in water, but calcium soaps are not.

27.   The nonpolar tail of the soap molecule dissolves in water, and the polar head of the molecule dissolves in grease.

28.   Hard water contains metal ions that form insoluble salts with soap.

29.   Water softeners work by removing metal ions from the water and replacing them with ions that do not interfere with the soap.

30.   Changing the pH of water will affect the cleansing action of detergents.

31.   The calcium salts of detergents are water soluble.

32.   Synthetic detergents have a molecular shape similar to that of fatty acid soaps.

To review: the five reactions of fats and oils of major interest are:

▸	Iodination	Gives the iodine number that is a measure of the degree of unsaturation.
▸	Hydrogenation	Converts carbon-to-carbon double bonds to single bonds and by that decreases the iodine number and increases the melting point.
▸	Hydrolysis	Produces glycerol and three molecules of fatty acid.
▸	Saponification	This reaction with strong base produces glycerol and three salts of fatty acids (soaps).
▸	Rancidity	Hydrolysis and oxidation produce strong odors and tastes due to the synthesis of carboxylic acids and aldehydes.

16.11 Compound Lipids

Compound lipids, upon hydrolysis, yield fatty acids, an alcohol and other compounds (such as phosphate or a sugar).

The phospholipids are compound lipids containing glycerol, two fatty acids, phosphoric acid, and a nitrogen compound. The structure of these compounds is similar to that of the soaps, and they have good emulsifying properties. They form the double-layer framework for cell membranes.

Phosphatidylcholine (PC or lecithin) is a phosphoglyceride in which the nitrogen compound is choline. PC is important in the metabolism of fats in the liver, is a source of phosphate for tissue formation, and helps in the transport of fats.

Sphingolipids are phospholipids that contain the alcohol sphingosine instead of glycerol. The most common sphingolipid is sphingomyelin, which forms part of the protective coating (called the myelin sheath) around nerve cells.

Glycolipids have a structure very similar to phospholipids, except that the phosphate group is replaced by a sugar that is usually galactose, but may also be glucose. Cerebrosides are glycolipids containing sphingosine, and are found in high concentrations in the myelin sheath.

Important Terms

phospholipid	phosphatidylcholine	lecithin
sphingolipid	sphingosine	myelin sheath
sphingomyelin	glycolipid	cerebroside

Check Your Understanding ───────────────────────────

For questions 33 to 42, match the statement with the correct answer or answers in column B. An item in column B may be used more than once.

		Column B
33.	Form the framework of cell membranes	A. wax
34.	Always contain glycerol	B. phospholipids
35.	Ester of a long-chain alcohol and a long-chain fatty acid	C. phosphatidylcholine
36.	Contains the alcohol sphingosine	D. sphingolipids
37.	Important emulsifying agent	E. myelin sheath
38.	Forms part of the myelin sheath	F. glycolipids

39. Similar to phospholipids but contains a sugar group rather than a phosphate group
40. Source of phosphate for tissue formation
41. Component of protective coating of skin, feathers, and leaves
42. Protective coating around nerve cells

16.12 Nonsaponifiable Lipids: Steroids

Steroids are a large class of compounds having a wide range of functions (see Table 16.5 in the text). The basic structure of all steroids is a nucleus containing three six-carbon rings and one five-carbon ring.

Sterols are steroid alcohols. Cholesterol is the most abundant sterol and is found in most animal tissue. Cholesterol is synthesized by the liver from acetyl-CoA, is used in the synthesis of bile salts, sex hormones, and vitamin D, and plays a role in the development of atherosclerosis.

Important Terms

steroid   sterol   cholesterol

16.13 Cellular Membranes

Cellular membranes contain lipids and proteins and small amounts of carbohydrate. The membrane structure is a lipid bilayer containing two rows of phospholipids with their polar heads to the outside and their nonpolar tails to the inside. The protein molecules are found spaced throughout the membrane (See Figure 16.5 in the text).

Check Your Understanding ─────────────────────────────

For questions 43 to 50, fill in the blanks with the correct word or words.
43. Vitamin D, cortisone, and progesterone are all members of the class of compounds called _____.
44. Steroids belong to the large class of lipids that are _____ (not hydrolyzed by base).
45. Steroids all share the same common _____.
46. Cortisone and testosterone are steroid _____.
47. Digitoxigenin is used as a drug to treat _____.
48. Sterols are steroids containing a(n) _____ group. The most abundant sterol is _____. This sterol is synthesized in the _____ from _____. It is carried in the blood in the form of _____, _____, and _____.
49. The two major components of cellular membranes are _____ and _____.
50. Describe the structure of a cellular membrane.

1. a, b **2.** b **3.** b, c **4.** a, d **5.** c **6.** a **7.** b **8.** d **9.** d **10.** a **11.** c **12.** a, c, d **13.** c **14.** d **15.** c **16.** b **17.** unsaturation **18.** oils, fats **19.** hydrogenation, solid (saturated), liquid (unsaturated) **20.** arachidic **21.** odor, taste, rancid **22.** glycerol and three molecules of the same fatty acid **23.** saponification, glycerol, fatty acid salts **24.** potassium hydroxide, sodium hydroxide

25. T **26.** F **27.** F **28.** T **29.** T **30.** F **31.** T **32.** T **33.** B, D **34.** B, C **35.** A **36.** D **37.** C **38.** D **39.** F **40.** C **41.** A **42.** E **43.** steroids **44.** nonsaponifiable **45.** four ring structure **46.** hormones **47.** heart disease **48.** hydroxyl (alcohol), cholesterol, liver, acetyl-CoA, chylomicrons, LDL, HDL **49.** lipids, proteins **50.** The cell membrane is a lipid bilayer with proteins and carbohydrate components imbedded in the bilayer.

Chapter 17   PROTEINS

This chapter discusses the most complex and varied molecules of life, the proteins. First we present several methods of classifying this large group of compounds. Then, we discuss the properties of amino acids, the monomers that form the polymer proteins. Proteins have complex structures, and their structure determines their functions. In the second section of this chapter, we discuss protein structure and the factors that affect this structure.

17.1   Amino Acids: Building Blocks of Proteins

Proteins are extremely important biological molecules, serving many functions in the cell. They are very large polymers of amino acids; their size is in the range of colloidal particles.

Amino acids contain both an amino and a carboxyl functional group. The amino group is always on the carbon next to the carboxylic acid group, this is the alpha carbon. Most naturally occurring proteins are composed of combinations of only 20 different amino acids, which can be identified by their R-group side chains.

Important Terms

      amino acid   protein

17.2   The L-Family of Amino Acids

Except glycine, all amino acids that are found in proteins are optically active and belong to the L-family.

Check Your Understanding ────────────────────────────

1.   Draw the structures for (a) L-phenylalanine and (b) L-serine.

17.3   Essential Amino Acids

Ten of the 20 amino acids found in proteins cannot be synthesized in sufficient amounts by the human body. These amino acids, called the essential amino acids, must be supplied in the diet. A protein that contains all 10 essential amino acids is called an adequate protein. Animal protein and milk are adequate proteins, but many vegetable proteins are not. Protein deficiency diseases are common in those parts of the world where a single plant, such as corn, is the main source of protein. Corn is too low in lysine and tryptophan to support growth in children. A combination of vegetables can supply all of the essential amino acids.

Important Terms

essential amino acid   adequate protein

17.4   Acid-Base Properties

When dissolved in water, amino acids exist as zwitterions, or dipolar ions. Amino acids are amphoteric: they can act either as an acid or a base. Proteins function as buffers in the blood.

Important Terms

amphoteric   zwitterion

17.5   Isoelectric Point

The isoelectric point is the pH at which an amino acid or protein is electrically neutral and will not migrate in an electric field. Each amino acid and each protein has a characteristic isoelectric point. Proteins are least soluble at their isoelectric points.

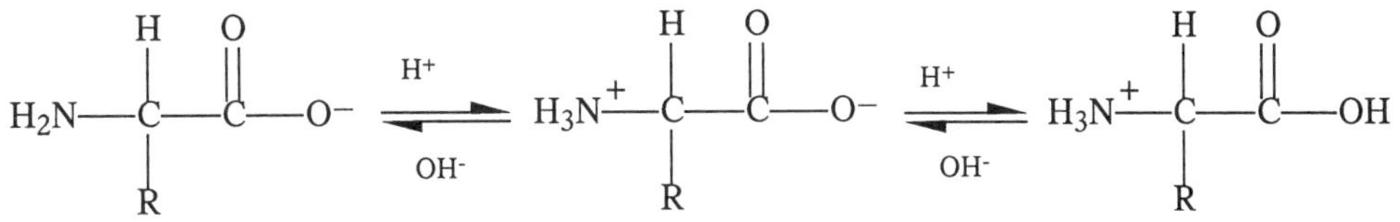

At a pH above (more basic) than its isoelectric point the amino acid exists as the anion. At a pH below (more acidic) than its isoelectric point the amino acid exists as a cation. At the isoelectric point the amino acid exists as the zwitterion, which is electrically neutral.

Important Term

isoelectric point (pI)

Example _____

Draw the structural formula of leucine in a solution of pH 10

S   Draw the structure of leucine in the form that it will exist in a solution of pH 10

T•E   Look up the R group for leucine in Table 17.1. It is $(CH_3)_2CHCH_2-$. Look up the isoelectric point for leucine in Table 17.3. It is 6.0. Therefore, leucine will exist in its anionic (negative) form.

P   The structure of leucine at pH 10 is:

$$\begin{array}{c} \qquad\qquad \overset{\displaystyle O}{\overset{\|}{}} \\ H_2N{-}CH{-}C{-}O^- \\ \quad | \\ H_2C{-\!-}CH{-}CH_3 \\ \qquad\qquad | \\ \qquad\qquad CH_3 \end{array}$$

Check Your Understanding _____

For questions 2 to 6, fill in the blanks with the correct word or words.

2. Amino acids are amphoteric, which means that they _____.
3. In water, amino acids exist as highly polar ions called _____.
4. Amino acids that the body cannot synthesize in sufficient amounts are called _____. The protein in beef is an example of an _____ protein, but the protein in rice is not because it does not contain sufficient amounts of _____ or _____.
5. The pH at which an amino acid or protein will not migrate in an electric field is called its _____. At a pH more basic than this point, a protein will carry a net _____ charge and will migrate toward the _____ pole in an electric field.
6. Proteins are the least soluble at a pH _____ (less than, equal to, or more than) their isoelectric points.

7. (a) Draw the structure of alanine in neutral water.
   (b) Show how alanine can act as a buffer and can protect against changes in pH when an acid or a base is added to the solution.

For questions 8 to 10 describe
   (a) The charge [positive, negative, no charge] the molecule will carry at pH 6.
   (b) Toward what pole [negative or positive] it will migrate if placed in an electric field pH 6.

8. alanine   9. egg albumin   10. hemoglobin

17.6 Proteins

Proteins may be classified as simple proteins containing only amino acids, or conjugated proteins containing other inorganic or organic components called prosthetic groups. Globular proteins are soluble in water, are quite sensitive to their chemical environment, and often function as enzymes. Fibrous proteins are insoluble in water, are fairly insensitive to their chemical environment, and have a structural or protective function. Proteins can also be categorized based on their functions in living organisms. Table 17.5 in the text lists some of these categories.

simple protein   globular protein   conjugated protein
fibrous protein   prosthetic group

Check Your Understanding _____

For questions 11 to 20, choose the best answer or answers.

11. Because of their size, proteins belong to which class of compounds?
     (a)   crystalloids   (c)   glycoproteins
     (b)   colloids   (d)   lipoproteins

12. Components of proteins that are not amino acids are called
     (a)   conjugated groups   (c)   prosthetic groups
     (b)   metal ion   (d)   enzymes

13. Proteins that function as structural components of organisms are called
     (a)   simple proteins   (c)   globular proteins
     (b)   conjugated proteins   (d)   fibrous protcins

14. The monomer units of proteins are
     (a)   sugars   (c)   enzymes
     (b)   amino acids   (d)   carboxylic acids

15. These proteins are soluble in water.
     (a)   simple proteins   (c)   globular proteins
     (b)   conjugated proteins   (d)   fibrous proteins

16. These proteins are physically tough.
     (a)   glycoproteins   (c)   globular proteins
     (b)   conjugated proteins   (d)   fibrous proteins

17. These proteins function as enzymes.
     (a)   glycoproteins   (c)   globular proteins
     (b)   contractile proteins   (d)   fibrous proteins

18. Proteins that contain only amino acids are called
     (a)   simple proteins   (c)   globular proteins
     (b)   conjugated proteins   (d)   fibrous proteins

19. Hemoglobin belongs to this class of proteins.
     (a)   structural protein   (c)   storage protein
     (b)   transport protein   (d)   protective protein

20. Antibodies belong to this class of proteins.
     (a)   structural protein   (c)   storage protein
     (b)   contractile protein   (d)   protective protein

17.7   Primary Structure

The primary structure of a protein is its sequence of amino acids. The amino acids are linked by peptide bonds, which are amide linkages formed by condensation reactions between the amino acids. The peptide bond is stable, and can be broken only by certain enzymes or by acid or base hydrolysis. A dipeptide contains two amino acids, a tripeptide contains three amino acids, an oligopeptide four to ten amino acids, and a polypeptide more than ten.

The amino acid sequence of a protein can be indicated using three-letter abbreviations for the names of the amino acids. Dashes or dots are used to show the peptide bonds. The end of the molecule containing the free amino group is called the N-terminal end, and is usually written first in the sequence. The end containing the free carboxyl group is called the C-terminal end, and is written last in the sequence.

The amino acid sequence is critical to the structure and function of the protein. A change in just one amino acid can totally disrupt the normal functioning of the protein.

Important Terms

primary structure	dipeptide	peptide bond
tripeptide	N-terminal end	oligopeptide
C-terminal end	polypeptide	

Example _____

What is the name of this tripeptide?

S   Determine the name of the tripeptide.

T•E   Identify the three "R" groups in the tripeptide and then name the compound starting at the N-terminal group. The R groups in order are: CH_3-, $CH(CH_3)_2$- and CH_3-
When R= CH_3, the amino acid is alanine.
When R = $CH(CH_3)_2$ the amino acid is valine.

P   The name of the tripeptide is alanylvalinylalanine, or Ala-Val-Ala in shorthand notation.

21. (a) Draw the structure of the tripeptide formed between alanine, serine, and valine. Put valine on the N-terminal end and serine on the C-terminal end and label the peptide bonds.

    (b) Write the three letter abbreviations for this sequence.

17.8 Secondary Structure

The secondary structure of proteins is the shape of specific regions of the polypeptide backbone. It results from hydrogen bonding between the amino acids, and takes on several configurations: the alpha helix, the beta configuration or pleated sheet, and the triple helix.

In the alpha helix the amino acids form loops, with the R-groups on the amino acids extending to the outside of the helix. In the beta configuration, several polypeptide chains are lined up next to one another, held together by hydrogen bonds. The R-groups on the amino acids extend above and below the plane of the sheet. The triple helix consists of three polypeptide chains twisted around each other to form a helix or spiral.

Each protein will have its own characteristic secondary structure determined by its sequence of amino acids. Some large globular proteins may have regions of alpha helix and regions of beta configuration in their structure. Unlike peptide bonds, hydrogen bonds are easily disrupted by changes in pH, temperature, solvents, or salt concentrations.

Important Terms

secondary structure	alpha helix	beta pleated sheet
keratin	beta configuration	collagen
triple helix	silks	

17.9 Tertiary Structure

Globular proteins have a tightly folded three-dimensional structure that results from disulfide bridges, hydrogen bonding, salt bridges, and hydrophobic interactions. This shape is the native state or native configuration of the protein molecule.

Disulfide bridges form between the side chains of two cysteine molecules, forming a covalent linkage between two regions on one polypeptide chain or between two polypeptide chains. These linkages are not sensitive to changes in pH, salt concentration, solvents, or temperature, and can be broken only by reduction.

Salt bridges are ionic attractions between the charged side chains of amino acids such as lysine or aspartic acid, and can be disrupted by changes in pH.

The hydrophobic side chains of certain amino acids congregate toward the inside of the molecule, away from the polar water molecules. These hydrophobic interactions are a key to the shape of the protein molecule.

Important Terms

    tertiary structure   native state   disulfide bridge
    hydrophobic interaction   salt bridge   hydrogen bonding

17.10 Quaternary Structure

The quaternary structure of a protein is the way in which two or more polypeptide chains are held together in a three-dimensional arrangement. Hydrogen bonding, salt bridges, and hydrophobic interactions may all be involved in holding the chains together.

Important Term

    quaternary structure

Check Your Understanding —————————————————————————

For questions 22 to 31, fill in the blanks with the correct word or words.

22.   The _____ structure of a protein is the sequence of amino acids in the protein. These amino acids are held together by _____ bonds that are quite stable and can be broken only by _____ or _____.

23.   The curled or spiral arrangement of amino acids in a polypeptide chain is called an _____. The loops are held in place by _____ between the atoms that are part of the peptide bonds.

24.   The zigzag arrangement of the polypeptides in silk is called the _____. It consists of several parallel polypeptide chains held together by _____.

25.   The geometric arrangements of the amino acids discussed in questions 23 and 24 form the _____ structure of a protein.

26.   The long polypeptide chains of globular proteins are tightly folded in an arrangement called the _____ structure of the protein. This folding is a result of interactions between the _____ of the amino acids and may involve one or more of the following interactions: _____, _____, _____, and _____.

27.   Quaternary structure occurs when the protein has more than one _____. The interactions involved in this structure may be _____, _____, or _____.

28.   Hydrogen bonding is a weak _____ (covalent, noncovalent) linkage, and is easily

disrupted by changes in _____, _____, _____ or _____.

29. _____ result when the sulfhydryl groups on the amino acid cysteine undergo oxidation. This is a _____ (covalent, noncovalent) linkage that can be disrupted by _____, but is stable during changes in _____, _____, or _____.

30. _____ form between the charged groups on the side chains of amino acids such aspartic acid, and are easily disrupted by changes in _____.

31. _____ form between the R-groups of such amino acids as valine or isoleucine, and occur on the _____ (inside, outside) of the protein molecule.

17.11 Denaturation

A protein can carry out its biological function only when it is in its native state. Denaturation is the disruption of the native state of the protein. Denaturation may or may not be permanent, depending upon the conditions. Changes in pH disrupt hydrogen bonding and salt bridges in proteins, and will cause denaturation. Heat and radiation will increase the motion of a protein molecule, disrupting hydrogen bonding and salt bridges. Small changes in temperature cause reversible denaturation, but high temperatures result in irreversible denaturation and the coagulation of protein. Organic solvents such as alcohols interfere with the hydrophobic interactions of the nonpolar side chains of the protein. Ions of heavy metals such as lead, mercury, or silver may disrupt salt bridges or disulfide bridges, and are toxic to organisms. Alkaloid reagents such as tannic acid affect salt bridges and hydrogen bonding. Reducing agents will break disulfide bridges in a protein.

Important Term

denaturation

Check Your Understanding _____

For questions 32 to 39, answer true (T) or false (F).

32. When a protein is denatured, the primary structure is disrupted.

33. Small changes in pH and in temperature will cause reversible denaturation of proteins.

34. Changes in pH affect disulfide bridges in a protein.

35. Strong heating of a protein will cause it to coagulate.

36. Heavy metal ions do their damage to the salt bridges in a protein.

37. Eggs are a good antidote for mercury poisoning.

38. Sunburn is a result of denaturation of proteins in the skin.

39. Reducing agents disrupt salt bridges.

1. (a)

$$H_2N-\underset{\underset{\displaystyle C_6H_5}{|}}{\overset{\overset{\displaystyle COOH}{|}}{\underset{\displaystyle CH_2}{\overset{|}{C}}}}-H$$

(b)

$$H_2N-\underset{\underset{\displaystyle CH_2OH}{|}}{\overset{\overset{\displaystyle COOH}{|}}{C}}-H$$

2. can act as an acid or a base **3.** zwitterions **4.** essential amino acids, adequate, lysine, threonine **5.** isoelectric point, negative, positive **6.** equal to

7. (a) $H_3\overset{+}{N}-\underset{\underset{\displaystyle CH_3}{|}}{CH}-COO-$

(b) $H_3\overset{+}{N}-\underset{\underset{\displaystyle CH_3}{|}}{CH}-COOH \xleftarrow{+H^+} H_3\overset{+}{N}-\underset{\underset{\displaystyle CH_3}{|}}{CH}-COO^- \xrightarrow{+OH^-} H_2N-\underset{\underset{\displaystyle CH_3}{|}}{CH}-COO^-$

8. (a) no charge (b) won't migrate **9.** (a) negative (b) positive **10.** (a) positive (b) negative **11.** b **12.** c **13.** d **14.** b **15.** c **16.** d **17.** c **18.** a **19.** b **20.** d

21. (a)

$$H_2N-\underset{\underset{\underset{\displaystyle CH_3}{|}}{\underset{\displaystyle CHCH_3}{|}}}{CH}-\overset{\overset{\displaystyle O}{\|}}{C}-NH-\underset{\underset{\displaystyle CH_3}{|}}{CH}-\overset{\overset{\displaystyle O}{\|}}{C}-NH-\underset{\underset{\displaystyle CH_2OH}{|}}{CH}-\overset{\overset{\displaystyle O}{\|}}{C}-OH$$

peptide bond

peptide bond

(b) Val-Ala-Ser

22. primary, peptide, enzymes, acid or base hydrolysis **23.** alpha helix, hydrogen bonds **24.** beta configuration (or beta pleated sheet), hydrogen bonds **25.** secondary **26.** tertiary, R-groups, hydrogen bonding, disulfide bridges, salt bridges, hydrophobic interactions **27.** polypeptide chain, hydrogen bonding, salt bridges, hydrophobic interactions **28.** noncovalent, pH, temperature, solvent, salt concentrations **29.** disulfide bridges, covalent, reducing agents, pH, temperature, salt concentrations **30.** salt bridges, pH **31.** hydrophobic interactions, inside **32.** F **33.** T **34.** F **35.** T **36.** T **37.** T **38.** T **39.** F

Chapter 18   ENZYMES, VITAMINS, AND HORMONES

A major function of globular proteins is to catalyze the reactions occurring in living organisms. Enzymes make up the largest and most highly specialized class of proteins. In this chapter we define the terms used to describe enzymes and their functions. We then discuss the method of enzyme action and the factors that affect this action. In the last section of this chapter, we study the ways in which enzyme activity is regulated and inhibited. Hormones play an important role in the regulation of enzyme activity in the body.

18.1   What is Metabolism?

Metabolism refers to all the enzyme-catalyzed reactions that occur in the body. These reactions include anabolic (or biosynthetic) reactions that synthesize molecules needed for repair and growth, and catabolic reactions that break down large molecules to form smaller products and cellular energy.

Important Terms

metabolism   catabolic reaction   anabolic reaction

18.2   Enzymes

Enzymes are globular proteins that catalyze the reactions of metabolism. Enzyme molecules are the largest and most highly specialized class of proteins. Because they can be used again and again, they are present in the cell in low concentrations.

The terms used to describe enzymes are defined at the end of this section in the text. Study this terminology carefully before continuing to read the chapter.

Important Terms

enzyme   cofactor   coenzyme
prosthetic group   proenzyme   substrate
active site   holoenzyme   apoenzyme
zymogen

Check Your Understanding ———————————————————————

For questions 1 and 2, fill in the blanks with the correct word or words.
1.   _____ is the term used to denote the chemical reactions occurring within the body.

2.	These reactions can be broken down into two types: _____ reactions that involve the breakdown of large molecules into smaller molecules and chemical energy, and _____ reactions that involve the synthesis of molecules needed by the cell.

For questions 3 to 11, match the statement in column A with the correct answer or answers from column B. An item in column B may be used more than once.

	Column A		Column B
3.	Tightly bound cofactor	A.	enzyme
4.	Protein portion of the enzyme	B.	coenzyme
5.	Apoenzyme + cofactor	C.	apoenzyme
6.	Inactive enzyme	D.	holoenzyme
7.	Enzyme acts on this substance	E.	prosthetic group
8.	Organic molecule as cofactor	F.	active site
9.	Protein catalyst	G.	substrate
10.	Chemical groups other than proteins required for enzyme activity	H.	cofactor
		I.	simple protein
11.	Region on enzyme to which substrate attaches	J.	conjugated protein
		K.	zymogen

18.3   Enzyme Nomenclature and Classification

As with the common names for organic compounds, the early nomenclature for enzymes was not at all systematic. These early names often gave no indication of the reaction being catalyzed nor of the substrate. In 1961, an enzyme classification system was proposed that named enzymes based on the substrate or type of reaction involved. This led to the establishment of six major divisions of enzymes, listed in Table 18.1 of the text.

Important Terms

hydrolases   oxidoreductases   transferases
lyases   isomerases   ligases

Check Your Understanding _____

For questions 12 to 17, choose the best answer.
12.	This class of enzymes catalyzes the interconversion of isomers.
(a)	hydrolases	(c)	lyases
(b)	transferases	(d)	isomerases
13.	This class of enzymes catalyzes oxidation-reduction reactions.
(a)	hydrolases	(c)	oxidoreductases
(b)	transferases	(d)	ligases

14. These enzymes catalyze hydrolysis reactions.
    (a) hydrolases   (c) lyases
    (b) isomerases   (d) ligases
15. With ATP, these enzymes catalyze the formation of new bonds.
    (a) oxidoreductases   (c) lyases
    (b) isomerases   (d) ligases
16. This class of enzymes catalyzes the transfer of functional groups.
    (a) hydrolases   (c) transferases
    (b) isomerases   (d) ligases
17. These enzymes catalyze the elimination of groups to form double bonds.
    (a) oxidoreductases   (c) lyases
    (b) isomerases   (d) ligases
18. What class of enzyme catalyzes the following reaction?

    glucose $\rightleftharpoons$ fructose

    (a) hydrolase   (c) lyase
    (b) transferase   (d) isomerase
19. What class of enzyme catalyzes the following reaction?

    glycylalanine + H_2O $\longrightarrow$ glycine + alanine

    (a) hydrolase   (c) lyase
    (b) transferase   (d) isomerase

18.4 Method of Enzyme Action

Enzymes lower the activation energy of a reaction by joining with the substrate and increasing the probability that a reaction will occur. (See the series of steps outlined in section 18.4 in the text.) Each enzyme catalyzes its reaction at a specific rate. The turnover number of an enzyme is the number of substrate molecules transformed per minute by one molecule of enzyme under optimum conditions.

Important Terms

enzyme-substrate complex   turnover number

18.5 Specificity

Enzymes differ from inorganic catalysts in their specificity. They might catalyze only one type of reaction or, in extreme cases, limit their activity to one specific compound.

One way to explain the specificity of enzymes is to picture the geometry of the active site as specially designed for a specific substrate. Other molecules won't fit in the active site, just as the shape of a lock will fit only one key, and other keys won't turn the lock. Not only must the geometry of the active site be correct, but the entire enzyme must be in its native state for it to function properly.

The lock-and-key model explains the action of some enzymes, but must be modified to fit the action of other enzymes. Enzyme molecules are flexible. Sometimes the active site may not fit the substrate, but may be induced by the substrate itself to take on the conformation of the substrate (just as a glove is induced by a hand to take on the conformation of the hand).

Enzymes that are secreted in the body as proenzymes are enzymes that have their active sites blocked. They become active when the site is unblocked—usually by the hydrolysis of some of the molecule. Cofactors function either to provide the active site for the enzyme or to form a bridge between the enzyme and the substrate.

Important Terms

      specificity   lock-and-key theory   induced-fit theory
      cofactor   proenzyme

Check Your Understanding   ——————————————————————

For questions 20 to 24, fill in the blank with the correct word or words.
20.   The four steps in an enzyme-catalyzed reaction are _____, _____, _____, and _____.

21.   Enzymes increase the rate of a chemical reaction by lowering the _____.
22.   The rate at which an enzyme catalyzes a reaction is described by its _____.
23.   If one molecule of the enzyme glutamic dehydrogenase will catalyze the transformation of 500 substrate molecules per second, then its turnover number is _____.

24.   One major difference between enzymes and inorganic catalysts is the enzyme's _____. This results from the fact that the enzyme has a specific area, called the _____, to which the substrate attaches.

18.6   Factors Affecting Enzyme Activity

For an enzyme to function properly, its structure must be in the native state. Changes in the chemical environment, such as changes in pH and temperature, will affect the secondary and tertiary structure of a protein. Each enzyme has an optimum pH at which its activity is greatest, and most body enzymes have maximum activity at normal body temperature. Temperatures and pH's that deviate from these optimum conditions will cause a decrease in the activity of the enzyme. Very high temperatures will permanently denature the enzyme.

Check Your Understanding   ——————————————————————

For questions 25 to 28, answer true (T) or false (F).
25.   The chemical environment of an enzyme is critical to its activity.

26. The digestive enzyme trypsin has its greatest activity in the stomach.
27. Enzymes are permanently denatured by both low and high temperatures.
28. The enzyme salivary amylase, found in saliva, has its greatest activity at a temperature above 80°C.

18.7 Inhibition of Enzyme Activity

Irreversible Inhibition
Irreversible inhibition of an enzyme occurs when a functional group in the active site or a cofactor required for the enzyme's activity is destroyed or permanently modified.

Reversible Inhibition
Competitive inhibition occurs when a compound, whose structure is similar to that of the substrate, competes with the substrate for the active site, thus decreasing the activity of the enzyme. Noncompetitive inhibition involves the inhibitor combining reversibly with a part of the enzyme (other than the active site) that is essential to the activity of the enzyme. Thus, the difference between competitive and noncompetitive inhibition is the place on the enzyme to which the inhibitor attaches. Competitive inhibition occurs at the active site and noncompetitive inhibition occurs at some other site on the enzyme.

Antimetabolites are a class of compounds that inhibit enzyme action. Antibiotics are antimetabolites that work by preventing the growth of microorganisms. Antibiotics function in many ways: inhibiting the formation of the bacterial cell wall, increasing the permeability of the cell membrane, or interfering with protein synthesis within the cell. Chemicals used in cancer chemotherapy are often antimetabolites that compete for the active site on an enzyme in the cancerous cell.

Important Terms

antimetabolite	antibiotic	chemotherapy
irreversible inhibition	reversible inhibition	
competitive inhibition	noncompetitive inhibition	

Check Your Understanding

For questions 29 to 33, fill in the blanks with the correct word or words.
29. When an inhibitor binds tightly to a metal ion cofactor, the inhibition is _____.
30. Inhibition caused by a molecule whose shape closely resembles that of the substrate is called _____ inhibition.
31. Inhibition caused by a compound that binds reversibly with the sulfhydryl groups on the enzyme is called _____ inhibition.

32.   An antimetabolite is a compound that _____.

33.   _____ are antimetabolites that are extracted from cultures of microorganisms.

For questions 34 to 39, match the statement in column A with the correct answer or answers from column B.

Column A	Column B
34. Inhibits cholinesterase	A. iodoacetic acid
35. Antimetabolite used in cancer therapy	B. lead ion
36. Binds irreversibly to sulfhydryl groups	C. 5-fluorouracil
37. Binds reversibly to sulfhydryl groups	D. nerve gas
38. Inhibits the formation of bacterial cell walls.	E. cephalosporin
39. Competes with succinic acid for the active site	F. sulfanilamide
on succinate dehydrogenase	G. mercury ion
	H. penicillin
	I. malonic acid

18.8   Regulatory Enzymes

Cells must have ways to control the activity of enzymes during the life of the cell. Most events in cellular metabolism require a series of reactions, called a multienzyme system, each of which is catalyzed by a specific enzyme. In a multienzyme system there is a regulatory or allosteric enzyme (usually the enzyme for the first reaction in the series) that controls the rate of the entire series. This enzyme is a complex molecule having an active site and one or more regulatory or allosteric sites. The activity of this enzyme is controlled (increased or inhibited) by the presence of substances in the cell called regulatory molecules.

Quite often these regulatory molecules are the end product of the reaction series. This type of regulation is called feedback control. In such a case, high concentrations of the end product will inhibit the regulatory enzyme. For example, high levels of ATP in the cell will inhibit the reactions that produce it. Another way for the cell to control the rates of reactions of the cell is to control the rates of production of the enzymes.

Important Terms

regulatory enzyme   regulatory site   allosteric site
regulatory molecule   multienzyme system   feedback control
allosteric enzyme

18.9   Hormones

Hormones are chemical substances that coordinate the actions between cells of a multicellular organism. Hormones are produced by a system of glands called the endocrine glands, and influence the activity of particular target organs or cells. The production of hormones may be triggered by the nervous system or by particular chemicals.

Cyclic AMP functions as a secondary messenger or intracellular hormone. Its formation from ATP is triggered by the attachment of a hormone to the membrane of a target cell. Cyclic AMP diffuses throughout the cell, causing the cell to respond to the hormone in a characteristic manner. For example, luteinizing hormone secreted by the pituitary travels through the blood and attaches to target cells in the ovaries, causing the production of cyclic AMP within the cells. The cyclic AMP then stimulates the production and secretion of progesterone by the cells.

Prostaglandins are 20-carbon fatty acids found in almost all cells. They are extremely potent biological agents and, like cyclic AMP, function to carry out messages that cells receive from hormones. Because of their potency, prostaglandins are finding increasing use as therapeutic agents.

Important Terms

hormone   endocrine gland   prostaglandin   cyclic AMP

Check Your Understanding ——————————————————————————————

For questions 40 to 51, fill in the blanks with the correct word or words.

40. A _____ is a sequence of enzyme-catalyzed reactions having a specific metabolic result.
41. The rate at which such a series of reactions occurs is controlled by a _____, or _____, enzyme.
42. This enzyme has a complex structure containing an _____ site and one or more _____ sites.
43. A _____ molecule attaches to this second site, causing a change in the enzyme's activity.
44. Many of these systems are controlled by the concentration of the _____ of the reaction. Such control is called _____.
45. Genes control the rate of chemical reactions within a cell by controlling the rate of enzyme _____.
46. In multicellular organisms, the activity of enzymes within cells is coordinated by _____ produced by the _____ glands.
47. The production of these chemical messengers is triggered by the _____ system or specific _____.
48. Hormones control the activity within the cell by _____ or _____.
49. _____ and _____ are produced in the cell and function as secondary messengers

within the cell. Their production is triggered by a _____ attaching to the cell membrane.

50. _____ is produced from ATP when the enzyme adenyl cyclase is activated.

51. _____ are potent chemicals that are used therapeutically to induce labor and relieve asthma.

18.10   What Are Vitamins?

Vitamins are organic compounds that the body cannot synthesize, but that are necessary for the normal growth and reproduction of cells. Most vitamins function as coenzymes, and the trace amounts we required must be supplied by the foods we eat. When a particular vitamin is missing in the diet, a specific deficiency disease results. Vitamins are classified as either water soluble or fat soluble.

Important Terms

vitamin	deficiency disease
coenzyme	RDA (Recommended Dietary Allowance)

18.11   Water-Soluble Vitamins

Any water-soluble vitamin absorbed in excess of the daily requirement is eliminated from the body in the urine. Therefore, these vitamins must be present in the diet every day. The water-soluble vitamins include vitamin C and the eight vitamins called the B vitamins. A deficiency in vitamin C causes scurvy; in thiamine, beriberi; in niacin, pellagra; and in vitamin B_{12}, pernicious anemia. Many of the B vitamins are coenzymes in reactions essential to life. Three of these coenzymes serve as hydrogen carriers in the oxidation-reduction reactions that supply cellular energy. Pantothenic acid forms part of the molecule coenzyme A.

Important Terms

water-soluble vitamin	vitamin C	thiamine (B_1)
riboflavin (B_2)	niacin	pyridoxine (B_6)
cyanocobalamin (B_{12})	pantothenic acid	folacin
biotin	scurvy	beriberi
pernicious anemia	pellagra	NAD^+
$NADP^+$	FAD	

18.12   Fat-Soluble Vitamins

Vitamins A, D, E, and K are fat-soluble. These vitamins are not soluble in water, and are stored in adipose tissue when absorbed into the body in excess of the daily requirements.

Although no harmful effects have been shown from high concentrations of water-soluble vitamins, absorption of large amounts of fat-soluble vitamins is very harmful. For example, vitamin D is necessary to prevent the deficiency disease rickets, but too much vitamin D is toxic.

Important Terms

vitamin A	vitamin D	vitamin E
vitamin K	rickets	

Check Your Understanding ─────────────────────────────────

For questions 52 to 65, choose the best answer or answers.

52. Many water soluble vitamins are
    (a)   stable   (c)   heat resistant
    (b)   coenzymes   (d)   stored in the body

53. Which of the following are fat-soluble vitamins?
    (a)   vitamin A   (c)   vitamin D
    (b)   vitamin B_1 (d)   vitamin E

54. Deficiency of this vitamin causes scurvy.
    (a)   vitamin A   (c)   vitamin C
    (b)   vitamin B_1 (d)   vitamin D

55. Low levels of this vitamin cause night blindness.
    (a)   vitamin A   (c)   thiamine
    (b)   niacin   (d)   vitamin K

56. When there is not enough of this vitamin in the diet the blood fails to coagulate.
    (a)   vitamin B_{12} (c)   vitamin E
    (b)   folacin   (d)   vitamin K

57. Which of the following are water-soluble vitamins?
    (a)   vitamin B_1 (c)   biotin
    (b)   niacin   (d)   vitamin E

58. This forms part of the coenzyme NAD^+.
    (a)   vitamin C   (c)   niacin
    (b)   riboflavin   (d)   pantothenic acid

59. Lack of this vitamin causes the deficiency disease pernicious anemia.
    (a)   vitamin B_1 (c)   vitamin B_6
    (b)   vitamin B_2 (d)   vitamin B_{12}

60. This vitamin is important in the regulation of calcium metabolism.
    (a)   vitamin D   (c)   pyridoxine
    (b)   vitamin E   (d)   thiamine

61. This vitamin forms part of coenzyme A.
    (a) vitamin C   (c) niacin
    (b) riboflavin   (d) pantothenic acid
62. Deficiency of this vitamin causes rickets.
    (a) vitamin A   (c) vitamin D
    (b) vitamin B_1 (d) vitamin E
63. Lack of this vitamin in the diet causes beriberi.
    (a) thiamine   (c) niacin
    (b) riboflavin   (d) pantothenic acid
64. A shortage of this vitamin causes pellagra.
    (a) thiamine   (c) riboflavin
    (b) niacin   (d) vitamin K
65. Any excess of these vitamins is stored in adipose tissue.
    (a) vitamin A   (c) vitamin D
    (b) vitamin C   (d) vitamin E

Answers to Check Your Understanding Questions in Chapter 18

1. metabolism **2.** catabolic, anabolic **3.** E **4.** C **5.** D, J **6.** K **7.** G **8.** B **9.** A **10.** H **11.** F
12. d **13.** c **14.** a **15.** d **16.** c **17.** c **18.** d **19.** a **20.** (a) The enzyme-substrate complex forms
(b) The substrate becomes activated (c) The products form on the surface of the enzyme (d)
The products are released from the surface of the enzyme **21.** activation energy **22.** turnover
number **23.** 3×10^4 molecules/min **24.** specificity, active site **25.** T **26.** F **27.** F **28.** F
29. irreversible **30.** competitive **31.** noncompetitive **32.** inhibits enzyme action
33. antibiotic **34.** D **35.** C **36.** G **37.** A **38.** H **39.** I **40.** multienzyme system
41. regulatory, allosteric **42.** active, regulatory (allosteric) **43.** regulatory **44.** end product,
feedback control **45.** synthesis **46.** hormones, endocrine **47.** nervous, chemical compounds
48. entering the cell and combining with a protein, attaching to the cell membrane
49. cyclic AMP, prostaglandins, hormone **50.** cyclic AMP **51.** prostaglandins **52.** b
53. a, c, d **54.** c **55.** a **56.** d **57.** a, b, c **58.** c **59.** d **60.** a **61.** d **62.** c **63.** a **64.** b **65.** a,c,d

Chapter 19   PATHWAYS OF METABOLISM

In this chapter we study the way in which the body digests carbohydrates, lipids, and proteins. We look at the cellular energy requirements, and how the cell uses molecules produced in the digestion of these food groups for synthesis of other compounds and for the production of energy.

19.1   Cellular Energy Requirements

The compounds used by the body to meet the immediate energy needs of cells are "high-energy" phosphates such as ATP. The hydrolysis of two oxygen-to-phosphorus bonds in ATP gives off large amounts of energy. This hydrolysis of ATP supplies the energy needed for the anabolic reactions in the cell.

Important Terms

> adenosine triphosphate, ATP   "high-energy" phosphate bond
> adenosine diphosphate, ADP   adenosine monophosphate, AMP

Check Your Understanding ——————————————————————

For questions 1 to 4, fill in the blank with the correct word or words.
1.   Compounds that contain _____ bonds supply the energy needed for endothermic reactions in the cell.
2.   One energy-rich compound is _____.
3.   In this compound, energy is released when a _____ -to- _____ bond is hydrolyzed.
4.   Hydrolysis of a high-energy phosphate bond releases _____ kcal of energy. An ordinary phosphate bond releases about _____ kcal.

19.2   Digestion and Absorption

Digestion is the process by which the body breaks down food so that it can be absorbed and used. Digestion of carbohydrates begins in the mouth, where enzymes in saliva begin the breakdown of starch. The remainder of carbohydrate digestion occurs in the small intestine, where enzymes in the pancreatic and intestinal juices break polysaccharides and disaccharides into the monosaccharides glucose, fructose, and galactose. These monosaccharides are absorbed through the intestinal wall into the blood and carried to the cells, where they enter various metabolic pathways.

The digestion of lipids begins in the stomach, where stomach acid begins the hydrolysis of fat. The majority of lipid digestion, however, takes place in the small intestine, where the bile salts emulsify the fats so they can be broken down by enzymes in the pancreatic and intestinal juices. Bile is a fluid manufactured by the liver and stored in the gall bladder before being released into the small intestine. It contains cholesterol, bile salts, and bile pigments. Fats are broken down to glycerol, fatty acids, and mono- and diacylglycerols, which cross the intestinal barrier and are carried to the blood via the lymph system. They are carried in the blood as micro-droplets joined to proteins called chylomicrons.

Digestion of proteins begins in the stomach, where enzymes activated by stomach acid break proteins down into polypeptides. The polypeptides are then hydrolyzed in the small intestine by enzymes in the pancreatic and intestinal juices to amino acids and small peptides. These products are absorbed into the bloodstream.

Important Terms

      digestion   bile   proenzyme

Check Your Understanding ————————————————————————

For questions 5 to 15, answer true (T) or false (F).
5. Many carbohydrates found in our food must be broken down into smaller molecules before they can be used by the cells in our body.
6. Digestion of carbohydrates begins in the stomach.
7. Lactose and maltose formed in the small intestine can be absorbed into the bloodstream.
8. Glucose formed in digestion may be stored in the liver as glycogen until needed.
9. Digestion of fats begins in the stomach.
10. Gall stones are crystallized bile.
11. Bile pigments are formed from the breakdown of hemoglobin.
12. Bile is produced and stored in the gall bladder.
13. Jaundice is caused by too much cholesterol in the blood.
14. Protein digestion begins in the mouth, where enzymes in saliva begin protein hydrolysis.
15. Protein-digesting enzymes are released in the stomach as proenzymes.

19.3 Carbohydrate Metabolism
Glycogenesis is the process by which glucose is polymerized to form glycogen in the liver when excess glucose is present in the blood. When the blood sugar level is low, glycogen is broken down in the liver by a series of reactions called glycogenolysis.

Insulin is produced by the beta cells in the pancreas when the concentration of glucose in the blood is high. It accelerates the absorption of glucose from the blood by muscle and fatty tissue, increases the production of glycogen, and decreases glycogenolysis in the liver.

The normal functioning of the brain requires a fairly constant level of glucose in the blood. Hypoglycemia is a condition caused by abnormally low concentrations of blood glucose. It can result in dizziness, lethargy, fainting, and (in severe cases) shock and coma. Hyperglycemia results from abnormally high concentrations of glucose in the blood, and occurs in conditions such as stress, kidney failure, and diabetes mellitus. When the glucose level exceeds that tolerated by the kidneys, glucose will enter the urine.

Important Terms

glycogenesis	glycogenolysis	hypoglycemia
hyperglycemia	insulin shock	insulin
glucagon	hyperinsulinism	renal threshold
glycosuria		

19.4   Cellular Respiration: Oxidation of Carbohydrates

The glucose in our blood enters the cells and is either used in biosynthetic reactions or is oxidized to carbon dioxide, water, and energy in the form of ATP in a process called cellular respiration. When we compare the energy produced by the oxidation of glucose in the laboratory with the energy produced by the oxidation of glucose in our cells, we find that our cells can trap about 44% of this energy; the rest of the energy is lost as heat to maintain body temperature.

The oxidation of glucose can be divided into two stages. The first stage, called glycolysis, is anaerobic (requires no oxygen.) It occurs in the cytoplasm of the cell, and produces two molecules of lactic acid and two molecules of ATP.

The second stage in the oxidation of glucose is aerobic (requires oxygen.) It occurs in the mitochondria of the cell, serves as the final stage of oxidation of other fuel molecules, and produces carbon dioxide, water, and ATP. In this stage, there are two series of reactions: the citric acid cycle and the electron transport chain.

Important Terms

cellular respiration	glycolysis	citric acid cycle
electron transport chain	respiratory chain	anaerobic
mitochondria	aerobic	

19.5   Citric Acid Cycle

The citric acid cycle is a cyclic series of reactions in which acetic acid (in the form of acetyl-CoA produced by the oxidation of pyruvic acid) is oxidized to two molecules of carbon dioxide, four pairs of hydrogen atoms, and one GTP. The hydrogen atoms are carried to the electron transport chain by NAD^+ and FAD. Keep in mind that the citric acid cycle cannot operate unless the electron transport chain is also functioning. The reactions of the citric acid cycle are shown in Figure 19.8 in the text, and you should be able to follow its cyclic nature. The first reaction is the coupling of acetyl-CoA with a molecule of oxaloacetate, and the last reaction produces a molecule of oxaloacetate to be used in another turn of the cycle.

Important Terms

coenzyme A   acetyl-CoA

19.6   Electron Transport Chain and ATP Formation

The four pairs of hydrogens produced in the oxidation of acetyl-CoA in the citric acid cycle enter the electron transport chain, where first these atoms, and then their electrons, are passed between a series of compounds. Each transfer involves an oxidation-reduction reaction in which energy is released. The exact mechanisms of the production of ATP in this process are not understood, but the points at which ATP is formed along the chain have been determined. Oxygen must be present for the electron transport chain to occur, and the products of this series of reactions are ATP and water. The citric acid cycle and the electron transport chain occur together, and anything that disrupts either series of reactions will disrupt the entire process of aerobic oxidation.

The oxidation of one molecule of glucose produces a total of 36 ATP:

- ▸   Glycolysis produces two ATP, two NADH and two pyruvate.
- ▸   Each pyruvate joins with acetyl-CoA producing one NADH.
- ▸   Each turn of the citric acid cycle produces three NADH, one $FADH_2$ and one ATP. There is one turn of the cycle for each pyruvate, therefore six NADH, two $FADH_2$ and two ATP are produced per glucose in the citric acid cycle.
- ▸   The processing of the NADH and $FADH_2$ in the electron transport chain results in the production of 32 ATP per glucose.

19.7   Lactic Acid Cycle

During strenuous exercise the body cannot supply the muscle cells with enough oxygen, so the cells must rely on glycolysis for the production of energy. Glycolysis can occur until the lactic acid that is produced hampers muscle performance, causing muscle fatigue and exhaustion. This lactic acid also enters the blood, causing acidosis and its complications. In order for the body to recover, the lactic acid must be oxidized by the muscle cells or liver, or be converted to glycogen by the liver. Heavy breathing after exercise supplies the oxygen necessary for these reactions.

Important Terms

lactic acid cycle   lactic acid   acidosis

19.8   Fermentation
Fermentation is an anaerobic process occurring in many microorganisms and in muscle cells. In fermentation, NAD$^+$ is regenerated by the transfer of electrons from NADH pyruvic acid or to a compound formed from pyruvic acid.

In alcohol fermentation, yeast cells obtain all the energy they need by glycolysis, but the end product of this glycolysis is alcohol rather than lactic acid.

Important Terms

fermentation   alcohol fermentation

Check Your Understanding ────────────────────────────────

For questions 16 to 23, fill in the blank with the correct word or words.
16.   The process of converting glucose to glycogen is called _____. The breakdown of glycogen to glucose is called _____. The rates of these two reactions series are controlled by the hormones _____ and _____.
17.   _____ controls the rate at which glucose in the blood enters the cells of the muscles and fatty tissue. It is synthesized by the _____ in the pancreas when the blood sugar level is _____ (high, low) and acts to _____ (lower, raise) the concentration of glucose in the blood.
18.   The normal fasting level of glucose in the blood of an adult is _____ per 100 mL of blood.
19.   A blood sugar concentration below normal is called _____. Its symptoms in mild cases are _____, and in severe cases _____.
20.   Overinjection of insulin results in a condition called _____, and can cause convulsions and coma referred to as _____.
19.   A meal high in _____ can cause overproduction of insulin and lowering of the blood sugar level.
22.   A high concentration of glucose in the blood is called _____ and can be caused by _____.

233

23.   When the level of glucose is above the _____, glucose will be excreted in the urine.

For questions 24 to 28, answer true (T) or false (F).
24.   Glycogenesis and glycolysis involve exactly opposite reaction pathways.
25.   Oxidation of glucose in cells is exothermic, and all of the energy is trapped in ATP molecules.
26.   The reactions of the citric acid cycle and the electron transport chain cannot occur separately.
27.   When sufficient oxygen is present, pyruvic acid is converted to acetic acid rather than lactic acid.
28.   The majority of ATP formed in the oxidation of glucose is produced in the electron transport chain.

For questions 29 to 45, choose the best answer or answers.
29.   Reactions that require oxygen are called
       (a)   anaerobic   (c)   anabolic
       (b)   aerobic   (d)   catabolic
30.   Another name for the Krebs cycle is the
       (a)   citric acid cycle   (c)   lactic acid cycle
       (b)   electron transport chain   (d)   fermentation
31.   A series of reactions in which the end products are water and ATP.
       (a)   citric acid cycle   (c)   lactic acid cycle
       (b)   electron transport chain   (d)   fermentation
32.   A substance produced in muscles during strenuous exercise.
       (a)   glucose   (c)   lactic acid
       (b)   glycogen   (d)   pyruvic acid
33.   The end products of this reaction are alcohol and carbon dioxide.
       (a)   glycogenesis   (c)   lactic acid cycle
       (b)   glycolysis   (d)   fermentation
34.   The enzymes of the citric acid cycle are found here.
       (a)   nucleus   (c)   chloroplast
       (b)   mitochondria   (d)   cytoplasm
35.   The process by which glycogen is synthesized from glucose.
       (a)   glycogenesis   (c)   glycolysis
       (b)   glycogenolysis   (d)   fermentation
36.   Reactions that can take place in the absence of oxygen are called
       (a)   anaerobic   (c)   anabolic
       (b)   aerobic   (d)   catabolic
37.   The series of reactions by which acetyl-CoA is oxidized to carbon dioxide.
       (a)   citric acid cycle   (c)   lactic acid cycle

(b) electron transport chain (d) fermentation

38. A series of redox reactions that release energy that is then trapped in molecules of ATP.
   (a) citric acid cycle (c) lactic acid cycle
   (b) electron transport chain (d) glycolysis

39. This compound causes muscle fatigue.
   (a) pyruvic acid (c) acetyl-CoA
   (b) citric acid (d) lactic acid

40. The reactions of glycolysis occur here.
   (a) nucleus (c) cytoplasm
   (b) mitochondria (d) chloroplast

41. The breakdown of glycogen to glucose occurs in
   (a) glycogenesis (c) glycolysis
   (b) glycogenolysis (d) fermentation

42. The series of reactions that involves the oxidation of glucose to lactic acid.
   (a) citric acid cycle (c) lactic acid cycle
   (b) electron transport chain (d) glycolysis

43. The molecule that carries acetic acid to the citric acid cycle.
   (a) pyruvic acid (c) NAD$^+$
   (b) coenzyme A (d) FAD

44. Hydrogen carriers in the electron transport chain.
   (a) pyruvic acid (c) NAD$^+$
   (b) coenzyme A (d) FAD

45. When glycolysis occurs in yeast it is called
   (a) glycogenesis (c) oxidation
   (b) glycogenolysis (d) fermentation

19.9 Lipid Metabolism

Adipose tissue contains triacylglycerols formed when excess lipids and carbohydrates are consumed. The synthesis of these lipids is called lipogenesis. Besides storing energy, adipose tissue functions as a support for inner organs and a heat insulator. Adipose tissue is not stable. It is constantly being synthesized and broken down as stored fatty acids are exchanged for food fatty acids.

Important Terms

lipogenesis adipose tissue

19.10 Oxidation of Fatty Acids

Fats yield more energy per gram than carbohydrates. Oxidation of triacylglycerols begins when they are hydrolyzed to glycerol and fatty acids. The glycerol enters the glycolysis pathway. The fatty acids are oxidized in a sequence of reactions called the fatty acid cycle or beta oxidation. These reactions (shown in Figure 19.12 in the text) are cyclic, and begin with the oxidation of the beta carbon. With each turn of the cycle, one acetyl-CoA and one pair of hydrogens are produced. The hydrogens enter the respiratory chain, and the acetyl-CoA's may enter the citric acid cycle or be used in the synthesis of other compounds.

Remember the six steps in the oxidation of fatty acids:
1.   The fatty acid bonds to coenzyme A. This step uses up an ATP
2.   Hydrogen is removed from the alpha and beta carbons. This forms a double bond between these two carbons. This is the reverse of the hydrogenation reaction seen in chapter 12. These hydrogens are carried to the electron transport chain by FAD.
3.   Water adds to the double bond. The hydroxyl group bonds to the beta carbon. This is the hydrolysis reaction seen in chapter 13.
4.   The secondary alcohol formed in step 3 is oxidized to a ketone. This is also a reaction seen in chapter 13.
5.   The bond between the alpha and beta carbons is broken. This reaction produces one molecule of acetyl coenzyme A and a fatty acid two carbons shorter than the process started with.
6.   The new fatty acid starts the process all over again at step 2.

Important Terms

beta oxidation   fatty acid cycle

19.11   Metabolic Uses of Acetyl Coenzyme A
Acetyl-CoA plays a central role in cellular metabolism. This is illustrated in Figure 19.13 in the text. Besides being the starting compound for the citric acid cycle, it is the starting material for the synthesis of compounds such as fatty acids, amino acids, and cholesterol. The body rids itself of excess acetyl-CoA by producing ketone bodies.

19.12   Ketone Bodies
When an abnormal increase in the level of fat metabolism occurs, excess acetyl-CoA is produced. This acetyl-CoA is converted in the liver to three compounds called ketone bodies. When the level of these compounds is high, ketosis results. Ketone bodies will be found in the urine and will cause acidosis in the blood. In extreme cases this can lead to coma and death.

Important Terms

ketone bodies   ketosis   acidosis

For questions 46 to 51, answer true (T) or false (F).

46.   Adipose tissue contains stored triacylglycerols.
47.   The composition of adipose tissue is determined by the type of diet.
48.   The oxidation of a gram of carbohydrate will yield about twice as much energy as a gram of fat.
49.   In the beta oxidation of fatty acids, four-carbon units are removed with each turn of the cycle.
50.   The only important role of acetyl-CoA in cellular metabolism is that it is the compound that enters the citric acid cycle.
51.   The formation of ketone bodies occurs when the rate of fat metabolism is reduced.

For questions 52 to 56, fill in the blank with the correct word or words.

52.   The oxidation of fats begins with the _____ of a triacylglycerol. The glycerol formed is oxidized in the _____ pathway, and the fatty acids are oxidized by the _____.
53.   In one turn of the fatty acid cycle _____, _____, and _____ are produced.
54.   The complete oxidation of behenic acid ($CH_3(CH_2)_{20}COOH$) will produce _____ ATPs.
55.   Ketone bodies are found in the urine in a condition called _____. In this condition, the pH of the blood will _____ (increase, decrease). This condition is caused by an increase in the level of _____ metabolism.
56.   An untreated diabetic can lapse into a coma caused by _____. This severe condition results because the diabetic must oxidize _____ to provide energy for the cells.

19.13 Protein Metabolism

The amino acids that enter the blood are used in the synthesis of new proteins and nonprotein, nitrogen-containing compounds. Amino acids that are absorbed in excess of the synthetic needs of the body are catabolized. The amino group is converted into urea and the remainder of the molecule enters the citric acid cycle or is used to synthesize other amino acids, glucose, and glycogen.

There is no storage form for amino acids in the body. Rather, the body maintains a pool of amino acids that constantly changes as protein is broken down and resynthesized.

Amino groups can be transferred from amino acids to form new amino acids in a process called transamination. In this series of reactions, an amino group is transferred from an amino acid to an α-keto acid, producing a new amino acid and a new α-keto acid.

Amino acids not used by the cell in synthetic reactions are catabolized to ammonia, carbon dioxide, water, and energy. The first step in this process is oxidative deamination, in which the amino group is removed from an amino acid to produce an α-keto acid and ammonia. The α-keto acid can enter the citric acid cycle, be used to produce other amino acids, or be used in the production of carbohydrates or fats. The ammonia produced enters the urea cycle.

To summarize: the three principle roles of amino acids that enter the blood stream are:
- Synthesis of proteins and other nitrogen-containing compounds.
- Formation of new amino acids through the process of transamination.
- Conversion to α-keto acids and ammonia by oxidative deamination.

Important Terms

transamination   α-keto acid   oxidative deamination

19.14 Urea Cycle
Ammonia is quite toxic to cells and cannot be allowed to build up within the body. Cells of the liver convert ammonia to the less toxic urea, which can be concentrated by the kidneys and excreted from the body. Figure 19.15 in the text shows the cyclic series of reactions of the urea cycle. Any disruption in the excretion of urea can be fatal.

Important Terms

urea   urea cycle   uremia

19.15 Pathways of Metabolism
The metabolism of carbohydrates, lipids, and proteins is intricately interrelated. This interrelationship is summarized in Figure 19.16. A disruption in any of these pathways will have profound effects on all aspects of metabolism within the human body.

Check Your Understanding _____

For questions 57 to 63, fill in the blanks with the correct word or words.
57.   Nitrogen used in synthesis reactions in the cells is obtained from _____ that we eat.
58.   The largest amounts of protein in the diet are needed during periods of _____ or _____.
59.   The body synthesizes new amino acids by a process called _____.
60.   In amino acid catabolism, amino acids are broken down first by a process called _____. This reaction produces an _____ acid that can enter the _____ cycle

or be used to synthesize_____, _____, or _____.

61. The other product of the reaction in question 60 is _____. It enters the _____ cycle, which takes place in cells of the _____.

62. A blockage of the kidneys can cause a build-up of urea and other wastes in the blood, resulting in a condition called _____. These wastes can be removed from the patient's blood by _____.

63. When dieting, it is important to maintain a certain minimum level of _____ in the diet to prevent the breakdown of tissue proteins and the formation of ketone bodies.

64. Write equations for the following reactions:
    (a) The transamination reaction between valine and α-ketoglutaric acid
    (b) The deamination of valine
    (c) The overall reaction of the urea cycle.

_____

Answers to Check Your Understanding Questions in Chapter 19

1. high energy phosphate **2.** ATP (UTP, GTP) **3.** oxygen, phosphorus **4.** 7, 2-5 **5.** T **6.** F **7.** F **8.** T **9.** T **10.** T **11.** T **12.** F **13.** F **14.** F **15.** T **16.** glycogenesis, glycogenolysis, insulin, glucagon **17.** insulin, beta cells, high, lower **18.** 60-100 mg **19.** hypoglycemia, irritability and grogginess, convulsions and coma **20.** hyperinsulinism, insulin shock **19.** refined carbohydrate **22.** hyperglycemia, diabetes mellitus, stress, kidney failure, certain drugs **23.** renal threshold **24.** F **25.** F **26.** T **27.** T **28.** T **29.** b **30.** a **31.** b **32.** c **33.** d **34.** b **35.** a **36.** a **37.** a **38.** b **39.** d **40.** c **41.** b **42.** d **43.** b **44.** c, d **45.** d **46.** T **47.** T **48.** F **49.** F **50.** F **51.** F **52.** hydrolysis, glycolysis, beta oxidation (fatty acid cycle) **53.** acetyl-CoA, NADH + H⁺, FADH₂ **54.** 181 **55.** ketosis, decrease, fat **56.** ketosis (acidosis), fats **57.** protein **58.** growth and cellular repair **59.** transamination **60.** oxidative deamination, α-keto, citric acid, amino acids, carbohydrates, fats **61.** ammonia, urea, liver **62.** uremia, hemodialysis **63.** carbohydrate

64. (a)

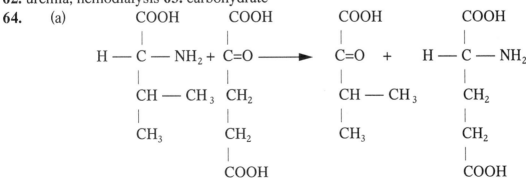

(b)

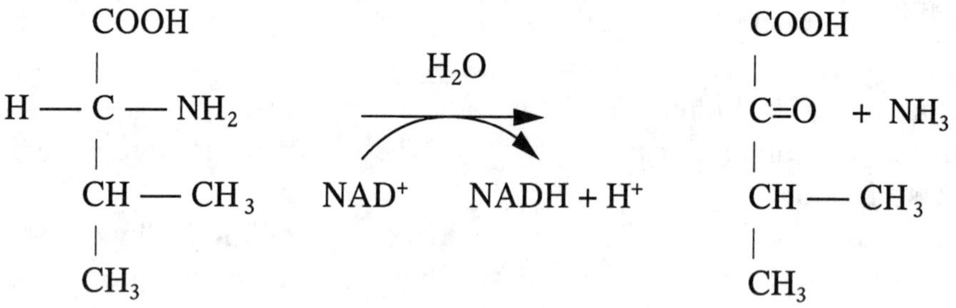
COOH
H — C — NH2
CH — CH3
CH3
H2O
NAD+
NADH + H+
COOH
C=O + NH3
CH — CH3
CH3

(c) $$2NH_3 + CO_2 \longrightarrow H_2N\text{-}\overset{\overset{\displaystyle O}{\|}}{C}\text{-}NH_2 + H_2O$$

Chapter 20   NUCLEIC ACIDS

Whether a cell will develop into a one-celled amoeba or a multibillion-celled human being is determined by the sequence of nitrogen-containing bases in a molecule called deoxyribonucleic acid, or DNA. This molecule belongs to a class of compounds called nucleic acids. In this chapter, we discuss the structure of nucleic acids, their method of replication, and the steps in protein synthesis. We see how errors in the DNA molecule can cause disease and death, and we discuss the molecular basis of the disease PKU.

20.1   Molecular Basis of Heredity

The specific characteristics of an organism are determined by its genes. Genes are segments of the DNA molecule that contain the information necessary to make one polypeptide chain. Nucleic acids, then, form the molecular basis of heredity because they control the nature of the proteins synthesized by the cell.

Important terms

        Central Dogma of Molecular Biology   gene

20.2   Nucleotides

DNA (deoxyribonucleic acid) and RNA (ribonucleic acid) are both nucleic acids, which are polymers of nucleotides. When hydrolyzed, nucleotides yield three components: a nitrogen-containing base, a five-carbon sugar, and phosphoric acid. The nitrogen bases are heterocyclic ring compounds that can be classified as pyrimidines or purines. The pyrimidines are uracil(U), thymine(T), and cytosine(C). The purines are adenine(A) and guanine(G). DNA contains the nitrogen bases: A, T, C, and G. RNA contains A, U, C, and G. Uracil is found only in the nucleotides of RNA, and thymine in the nucleotides of DNA. DNA contains the sugar deoxyribose, and RNA the sugar ribose.

		DNA	RNA
		phosphate	phosphate
Sugar:		2-deoxyribose	ribose
Bases:	purines	adenine and guanine	adenine and guanine
	pyrimidines	thymine and cytosine	uracil and cytosine

Besides forming the building blocks of nucleic acids, nucleotides serve as coenzymes and as high-energy phosphate molecules. The arrangement of nucleotides in a DNA molecule determines the genetic information carried by that molecule.

nucleic acid   nucleotide   pyrimidine
purine   uracil (U)   thymine (T)
cytosine (C)   guanine (G)   adenine (A)
ribose   deoxyribose   DNA
RNA

20.3   The Structure of DNA

In 1953 Watson and Crick described a molecule of DNA as two helical polynucleotide chains coiled around a common axis, with the nitrogen-containing bases toward the inside of the helix and the deoxyribose and the phosphate groups on the outside of the helix.

The backbone of the helix is a chain of deoxyribose sugars and phosphate groups connected by ester bonds as shown in Figure 20.1 in the text. The base on each nucleotide extends toward the inside of the double helix and is joined by hydrogen bonding to a base on the other strand. The pairing is very specific: adenine(A) pairs only with thymine(T), and guanine(G) only with cytosine(C). This means that the two strands of DNA are not identical, but rather are complementary. If we know the sequence of bases on one strand of DNA, base pairing allows us to determine the sequence on the second strand. Where there is an A on one strand there is a T on the complementary strand and vice versa. Where there is a G on one strand, there is a C on the complementary strand and vice versa.

Important Term

double helix

20.4   Replication of DNA

For genetic information to be passed on when a cell divides, the DNA molecules in the cell's nucleus must make identical copies of themselves—they must replicate. In this process, the two strands of the helix unwind and each serves as a template for the formation of a new strand. Each of the new double helixes produced will have one original DNA strand and one newly-made strand.

The genetic information carried by the DNA molecule is contained in the sequence of nitrogen bases in the molecule. A change in this order of bases will result in a mutation in the genes in the cell. The enzymes that replicate DNA have elaborate control mechanisms to check for and repair any errors in the newly synthesized DNA chains.

Important Terms

    replication   mutation

Check Your Understanding ─────────────────────────────────────

For questions 1 to 13, fill in the blank with the correct word or words.

1.   Genes are segments of molecules of _____, and carry the information required to produce one _____ chain.

2.   _____ are polymers of nucleotides.

3.   A nucleotide is composed of three subunits: _____, _____, and _____.

4.   Thymine and cytosine are in the class of bases called _____, and adenine and guanine are in the class of bases called _____.

5.   RNA contains the sugar _____, and DNA contains the sugar _____.

6.   A DNA molecule contains _____ polynucleotide chains, which form a _____.

7.   The bases in a DNA molecule are directed toward the _____ of the molecule. The _____ and _____ make up the backbone of the helix.

8.   The strands of the DNA molecule are not identical, but are _____.

9.   The bases of the two strands of DNA are joined by _____; the base cytosine will join only with _____, and adenine only with _____.

10.   When a DNA molecule produces another molecule identical to itself, it is undergoing the process called _____.

11.   In this process, the two strands of the DNA _____, and each strand serves as a _____ for the production of a new strand. Each daughter helix will contain one _____ strand and one _____ strand.

12.   The genetic information is contained in the _____ on the molecule of DNA.

13.   Anything that causes a change in this sequence of bases causes a _____ in the gene.

14.   (a)   Draw the structure of the nucleotide containing the base adenine and the sugar deoxyribose.

      (b)   Draw the structure of the nucleotide containing the base uracil and the sugar ribose.

      (c)   Would the nucleotide in (a) or (b) be found in RNA?

15.   The following is a sequence of bases found on one strand of a DNA molecule. What would be the sequence of bases on the other DNA strand of the helix?

                    A-G-G-T-A-C-G-A-C

16.   List three functions of nucleotides in the cell.

20.5   Ribonucleic Acid

RNA consists of a single strand of nucleic acid containing the sugar ribose and the bases adenine, cytosine, guanine, and uracil.

Messenger RNA

mRNA is synthesized on the DNA strand in the nucleus of the cell in a process called transcription. The mRNA has a sequence of bases complementary to the sequence on the DNA strand, except that uracil is found in RNA rather than thymine. mRNA then migrates to the ribosomes in the cytoplasm where it serves as a template for protein synthesis.

Ribosomal RNA

rRNA is synthesized on a DNA template in the nucleoli of the nucleus. rRNA joins with protein to form the two subunits of the ribosomes, the site of protein synthesis.

Transfer RNA

tRNA, synthesized in the nucleus on DNA, is the smallest and most mobile of the RNAs. Each tRNA is designed to attached to a specific amino acid, by that becoming "charged." This tRNA-amino acid complex migrates to the ribosomes where the amino acid is added to the end of the growing polypeptide chain.

Important Terms

messenger RNA (mRNA)	transcription	ribosomes
ribosomal RNA (rRNA)	transfer RNA (tRNA)	

Check Your Understanding _____

For questions 17 to 26, choose the best answer or answers.

17.   This molecule joins with proteins to form the ribosomes.
    (a)   DNA   (c)   rRNA
    (b)   mRNA   (d)   tRNA

18.   This molecule serves as a template for the sequence of amino acids in protein synthesis.
    (a)   DNA   (c)   rRNA
    (b)   mRNA   (d)   tRNA

19.   This molecule carries amino acids to the ribosomes.
    (a)   DNA   (c)   rRNA
    (b)   mRNA   (d)   tRNA

20.   If an amino acid attaches to the terminal adenine nucleotide, the result is a
    (a)   replicated DNA   (c)   charged tRNA

(b)   translated RNA   (d)   mutated DNA

21.   The site of protein synthesis in the cell is the
        (a)   ribosome   (c)   nucleus
        (b)   nucleolus   (d)   cytoplasm

20.   The smallest RNA molecule is
        (a)   mRNA   (c)   tRNA
        (b)   rRNA   (d)   hnRNA

23.   The first step in transmission of genetic information from a gene to a polypeptide
        chain is
        (a)   replication of DNA   (c)   charging of tRNA
        (b)   translation of mRNA   (d)   transcription of DNA to mRNA

24.   If the DNA template has the base sequence, A-T-A-T-C-C-G, it will produce an
        mRNA with the sequence:
        (a)   C-A-C-A-G-G-C   (c)   U-A-U-A-G-G-C
        (b)   T-A-T-A-G-G-C   (d)   A-T-A-T-C-C-G

25.   The nitrogen base found in RNA but not in DNA is
        (a)   adenine   (c)   guanine
        (b)   cytosine   (d)   uracil

26.   Which of these is not a component of a deoxyribonucleotide?
        (a)   adenine   (c)   guanine
        (b)   ribose   (d)   phosphate

27.   What would be the sequence of bases on a mRNA molecule synthesized on the DNA
        template shown below?

                                A-A-T-G-A-G-C-C-T

20.6 The Genetic Code

The genetic code is the term used to describe the specific sequence of bases found in a gene.
A gene is a segment of a DNA molecule that determines the sequence of amino acids in one
polypeptide chain. A sequence of three bases on mRNA, called a codon, is necessary to code
for each amino acid. Table 20.1 in the text shows these codons. Three codons do not code
for any amino acid, but instead are terminal codons signaling the end of the polypeptide
chain.

An anticodon is a three-base sequence on transfer RNA that is complementary to the codon
on the messenger RNA. Each tRNA molecule contains a region that carries the anticodon.

Important Terms

        codon   terminal codon   anticodon   human genome

245

28. The following are segments of three mRNA molecules. Mark off the codons on each mRNA and determine the sequence of amino acids for which it codes.

    (a)   U-U-C-U-C-A-A-C-U-G-A-U
    (b)   G-U-U-G-C-A-A-G-A-A-A-A
    (c)   A-U-C-C-C-A-C-A-C-C-U-G

20.7   Introns and Exons

Until recently, all of the scientific information on the transcription and translation of genetic information had come from studying one-celled bacteria having no nucleus. In these cells, mRNA is synthesized on a single strand of DNA and then combines with the ribosomes to synthesize proteins.

However, it has been discovered that in eukaryotic cells (cells with nuclei) the RNA synthesized on the DNA strand is much longer than the mRNA that exits the nucleus. This RNA, called heterogeneous nuclear RNA (hnRNA), contains sequences of bases called exons that code for amino acids, and intervening sequences of bases called introns. After the hnRNA is synthesized on the DNA, special enzymes in the nucleus cut the RNA and splice out all of the introns, producing the mRNA that migrates from the nucleus to the ribosomes in the cytoplasm.

Important Terms

heterogeneous nuclear RNA (hnRNA)   exon   intron

20.8   The Steps in Protein Synthesis

The first step in protein synthesis is the synthesis of a strand of mRNA—the transcription of DNA to mRNA. Next occurs the translation of the mRNA to protein. This sequence of reactions involves the attachment of amino acids to tRNA, the starting of a polypeptide chain, the addition of amino acids to the polypeptide chain, and the termination of the polypeptide chain. Many different ribosomes can be synthesizing protein on the same strand of mRNA. Such groups of ribosomes are called polysomes.

Important Terms

transcription of DNA   translation of mRNA   polysome

For questions 29 to 33, fill in the blanks with the correct word or words.

29.   The synthesis of mRNA on DNA is called _____, and the synthesis of a protein on the mRNA is called _____.

30.   In eukaryotic cells, _____ RNA, which is much longer than mRNA, is synthesized on the DNA. Special enzymes in the nucleus splice out the _____, which are sequences of bases that are between the _____, the sequences of bases that code for amino acids.

31.   Before the synthesis of a protein can begin, _____ must be synthesized in the nucleus, and tRNA must become _____.

32.   To begin the synthesis of a protein, the _____ subunit of a ribosome attaches to mRNA, and a tRNA with a specific initiating amino acid. This complex then joins with the _____ subunit to form the ribosome.

33.   The addition of amino acids to the polypeptide chain stops when a _____ is reached on the mRNA.

20.9   Genetic Engineering

Genetic engineering refers to those techniques by which genetic information from one species can be transferred to another species. When new DNA is inserted as part of an organism's own DNA the result is called recombinant DNA. Specific segments of DNA are spliced onto the DNA of *E. coli*, which then produces the desired protein, such as human insulin. Gene cloning is the process of making large quantities of a desired gene.

Important Terms

> gene cloning   human gene therapy   genetic engineering
> plasmid   recombinant DNA

20.10 Mutations

A mutation is a change in the sequence of bases on the DNA molecule. Such changes can occur spontaneously, or can be caused by viruses, radiation, or chemicals. The abnormal proteins that result from mutations can cause disorders, disease, or death to the organism. Over 3,500 diseases, such as sickle cell anemia, hemophilia, and PKU, result from mutations in genes and the defective enzymes produced by these mutations.

Important Terms

> mutation   genetic disorders

20.11 Phenylketonuria

Phenylketonuria (PKU) is a disease resulting from a mutation in the gene that codes for the liver enzyme phenylalanine hydroxylase. Phenylalanine hydroxylase normally catalyzes the conversion of the amino acid phenylalanine to the amino acid tyrosine. The defective enzyme produced in individuals inheriting PKU genes from both parents will not catalyze this reaction. As a result, phenylalanine and PKU metabolites, substances the body produces from phenylalanine in an attempt to rid itself of the excess, build up in the body. This produces an abnormal chemical environment for body cells, causing irreversible damage when PKU is untreated.

Important Terms

phenylketonuria, PKU   PKU metabolites

Check Your Understanding ————————————————————————————

For questions 34 to 42, match the statement in column A with the correct answer or answers in column B. An item in column B may be used more than once.

Column A		Column B
34.	This enzyme is defective in PKU	A. transcription
35.	PKU metabolite	B. translation
36.	A change in the base sequence of DNA	C. mutation
37.	Disease producing an accumulation of phenylalanine	D. phenylacetic acid
		E. recombinant DNA
38.	DNA $\longrightarrow$ mRNA	F. phenylalanine hydroxylase
39.	mRNA $\longrightarrow$ protein	G. PKU
40.	Phenylalanine $\longrightarrow$ tyrosine	
41.	T-C-A-G-C-A $\xrightarrow{\text{UV light}}$ T-C-T-G-C-A	
42.	Uses *E.coli* to produce human protein	

Answers to Check Your Understanding Questions in Chapter 20

1. DNA, polypeptide **2.** nucleic acid **3.** nitrogen-containing base, sugar, phosphoric acid
4. pyrimidines, purines **5.** ribose, deoxyribose **6.** two, double helix **7.** inside, sugar,

phosphate **8.** complementary **9.** hydrogen bonds, guanine, thymine **10.** replication
11. separate, template, original, new **12.** sequence of bases **13.** mutation
14. (a)   (b)   (c) b

15. T-C-C-A-T-G-C-T-G **16.** (a) coenzymes (b) energy-rich molecules (c) monomers for
nucleic acids **17.** c **18.** b **19.** d **20.** c **21.** a **22.** c **23.** d **24.** c **25.** d **26.** b
27. U-U-A-C-U-C-G-G-A
28. (a) UUC UCA ACU GAU (b) GUU GCA AGA AAA (c) AUC CCA CAC CUG
          Phe-Ser-Thr-Asp   Val-Ala-Arg-Lys   Ile-Pro-His-Leu
29. transcription of DNA, translation of RNA **30.** heterogeneous nuclear, introns, exons
31. mRNA, charged by attaching to an amino acid **32.** smaller, larger **33.** terminal codon
34. F **35.** D **36.** C **37.** G **38.** A **39.** B **40.** F **41.** C **42.** E

Chapter 21   BODY FLUIDS

In this chapter we study two extracellular fluids: blood and urine. Throughout this chapter you will find references to topics, such as acid-base chemistry, covered in earlier chapters. Refer to these chapters for review if you find that you have forgotten certain concepts.

21.1   Body Fluids

Some organisms have an open circulatory system. Fluid is pumped from the heart into the body cavity, where it diffuses into the tissues. Nutrients and waste are exchanged and then the fluid is pumped out of the body cavity and back to the heart. Vertebrates have a closed circulatory system. Blood is pumped through a series of large diameter blood vessels into increasingly branched and smaller vessels until it reaches the capillaries. As blood flows through the capillaries, nutrients and waste are exchanged between the blood and the cells.

Fluids within the cells are called intracellular fluids, while those outside the cells are called extracellular fluids. The extracellular fluids include: blood, urine, lymph, digestive juices, synovial fluid and interstititial fluid.

Important Terms

open circulatory system	closed circulatory system	blood
intracellular fluids	extracellular fluids	urine

21.2   Blood Components

Blood consists of blood plasma and suspended particles. The suspended particles, called the formed elements; are the red blood cells (erythrocytes), white blood cells (leukocytes) and the platelets. Blood plasma is about 92% water, 7% plasma proteins and 1% other dissolved material, such as inorganic salts and amino acids. The plasma proteins are: albumins, globulins and fibrinogen

If freshly drawn blood is allowed to stand, a clot will form. The yellow liquid that remains is blood serum. Serum is essentially the same as blood plasma except that it does not contain fibrinogen.

Important Terms

formed elements	hematocrit	blood plasma
blood serum	albumins	globulins
fibrinogens	erythrocytes	leucocytes

21.3   Functions of Blood

Seven functions of blood are listed in the text:

1.   Transport of oxygen and carbon dioxide. Refer to chapter 19 on metabolism
2.   Maintenance of fluid balance. Osmosis is covered in section 10.8
3.   Maintenance of proper pH. Acids and bases are covered in chapter 11 and equilibrium is covered in chapter 8
4.   Formation of clots.
5.   Transport of food and wastes.
6.   Regulation of body temperature. Specific heat is covered in chapter 2.
7.   Fighting invading microorganisms.

Check Your Understanding ———————————————————————————————

For questions 1 to 10 fill in the blanks.

1.   Single-celled organisms exchange nutrients and wastes with their surroundings through_____.
2.   Mollusks and snails have an _____ circulatory system. Fluid is pumped from the heart through a series of tubes into an _____ cavity in the body tissues.
3.   The circulatory system in vertebrates is a _____ system.
4.   Fluids that are not in the cells are called the _____ fluids. Examples of these fluids include_____, _____, _____ and _____.
5.   If fresh blood is centrifuged, the suspended material separates to the bottom of the tube. This material makes up about ____% of the blood and is referred to as the _____. The exact percentage is called the _____.
6.   The plasma proteins are: _____, _____ and _____.
7.   Blood plasma is approximately _____% water and _____% protein.
8.   Approximately 54 to 58% of the proteins in blood are _____.
9.   The _____ help maintain osmotic pressure.
10.   One of the functions of blood is the transport of the gases _____ and _____ between the lungs and the tissue.

21.4   Blood Gas Transport

The transport of oxygen and carbon dioxide is an important role of blood. Oxygen has a very low solubility in water. It is therefore transported primarily as a complex with hemoglobin(HbO_2). Carbon dioxide is transported primarily as bicarbonate, HCO_3^- (70%), but some is transported as dissolved carbon dioxide (10%) and some as a carbamino complex with hemoglobin ($HbNHCO_2H$) (20%).

The transport of oxygen and carbon dioxide in the blood involves a series of equilibria.

These are referred to as Equilibria I, II, and III in the text. Review Le Chatelier's principle and then go through the equilibria one by one studying the effect of increased oxygen and then increased carbon dioxide on the series of reactions. The figures in this section of the text should help you to see the relationships between these equilibria.

Example

What effect will an increase in the partial pressure of carbon dioxide have on the formation of oxyhemoglobin?

S   The question asks us to determine what effect an increase in the CO_2 partial pressure will have on the equilibrium for the formation of oxyhemoglobin.

T•E   We must use the equilibria for the reactions of carbon dioxide and determine what effect an increase in CO_2 partial pressure will have on these equilibria. These are Equilibria II and III. Then we can determine the effect of shifts in these two equilibria on the equilibrium for the formation of oxyhemoglobin-Equilibrium I.

Equilibrium II
$$H^+ + HCO_3^- \rightleftharpoons H_2CO_3 \rightleftharpoons H_2O + CO_2(g)$$

Equilibrium III
$$HbNH_2 + CO_2 \rightleftharpoons HbNHCO_2H$$

An increase in the partial pressure of CO_2 will shift equilibrium II to the left and Equilibrium III to the right. Hemoglobin will be used and protons will be released.

The equilibrium for the formation of oxyhemoglobin is:

Equilibrium I
$$HHb + O_2 \rightleftharpoons H^+ + HBO_2$$

P   The increase in the concentration of protons from equilibrium II and the decrease in hemoglobin will both tend to shift equilibrium I to the left. The formation of oxyhemoglobin will decrease.

Important terms

Hemoglobin   carbamino complex   oxyhemoglobin

21.5   Fluid Balance

Osmotic pressure occurs because of concentration differences across differentially permeable membranes. The solution with the higher concentration has higher osmotic pressure. Although solvent will move in both directions through the membrane, the net flow of solvent is from the side of the membrane with lower osmotic pressure(osmolarity) to the side with higher osmotic pressure. If osmosis continues, the osmotic pressures on both sides of the membrane will become equal.

The regulation of blood osmotic pressure is important in the maintenance of fluid balance in the body. Electrolytes are only partially stopped from moving through the walls of the capillaries. Therefore, a difference in electrolyte concentration will cause only a temporary difference in osmotic pressure. Because of their larger size, the plasma proteins cannot pass through the pores in the walls of the capillaries. It is the plasma proteins, and particularly the albumins, which maintain the osmotic pressure difference across the capillary walls.

Fluid balance in the body is subject to changes in blood osmotic pressure as well as the physical blood pressure. Fluid build up in the tissues is called edema, while excessive fluid loss leads to dehydration. Study the figures in section 21.5 to ensure that you understand this delicate balance.

Important Terms

osmotic pressure   blood pressure   edema

21.6   Regulation of Blood pH

The pH of human blood is normally about 7.4. Small changes in blood pH cause considerable distress, with death usually resulting if the blood pH rises above 7.8 or below 7.0.

Acidosis is the term used for low blood pH.   Acidosis can be due to metabolic or respiratory factors. Any condition in which there is an excessive intake of acidic substances or a restriction in the removal of acidic substances can cause acidosis. If the exchange of blood gases is reduced due to hypoventilation, carbon dioxide will be retained and blood pH will fall. Other causes of acidosis include ingestion of excessive amounts of acid, severe diarrhea, diabetes and failure of the kidneys to remove protons.

Alkalosis refers to high blood pH. The causes of alkalosis are just the opposite of those for acidosis: removal of too much acid or excessive intake of basic substances. Hyperventilation causes an excessive loss of carbon dioxide and blood pH rises. Severe vomiting causes a loss of acid and the consequent alkalosis. Carefully study Table 21.2. You should be able to predict the effects on blood pH of each of the causes listed in the table.

In chapter 11, we saw that buffers can be used in the lab to help maintain relatively constant solution pH. The buffers in our blood do the same thing. They help to maintain blood pH at close to 7.4. The major blood buffer is the carbonic acid-bicarbonate system. The phosphate buffer system works mainly within cells.

Let us see how Le Chatelier's principle is applied to the carbonic acid-bicarbonate buffer system.

$$H_2CO_3 \rightleftharpoons HCO_3^- + H^+$$

If acid is added to this buffer system, it will shift to the left to use up most of the excess acid.

$$H_2CO_3 \longleftarrow HCO_3^- + H^+$$

If base is added to this buffer system, it will react with the protons. The equilibrium will shift to the right to provide more protons to react with the base.

$$H_2CO_3 \longrightarrow HCO_3^- + H^+$$

Example
_____

Explain what effect starvation will have on the carbonic acid-bicarbonate buffer system.

S The question asks us to predict how the carbonic acid-bicarbonate system will shift to compensate for pH changes caused by starvation.

T•E As we saw in chapter 19, during starvation, the body's carbohydrate reserves are used up very quickly and lipid metabolism must be used to provide energy. This causes a build up of ketone bodies in the blood (ketosis) and a drop in blood pH.

P To compensate for a decrease in blood pH the carbonic acid-bicarbonate buffer system will shift to the left to try to remove excessive protons.

$$H_2CO_3 \longleftarrow HCO_3^- + H^+$$

Note that there is a limit to the ability of any buffer system to prevent pH changes. If too much acid or base is added, the components of the buffer system will all be used up.

Important terms

buffer   acidosis   alkalosis
bicarbonate buffer   phosphate buffer

Check Your Understanding _____

For questions 11 to 20 fill in the blanks.

11. The solubility of oxygen in blood is very _____(high, low).

12. Most of the oxygen is transported by the blood as _____.

13. _____% of the carbon dioxide transported by the blood is in the form of the dissolved gas, 20% is transported as a complex with _____, and the other _____% is in transported as _____.

14. In the tissues, the partial pressure of carbon dioxide is _____(high, low) and the partial pressure of oxygen is _____(high, low).

15. An increase in the partial pressure of carbon dioxide causes a(n) _____ (increase, decrease) in its solubility in blood.

16. When solutions of different concentrations are separated by _____, an osmotic pressure difference occurs.

17. The _____ make up more than half of the plasma proteins, so they have the greatest effect on blood _____ pressure.

18. Hypoventilation will cause the pH of blood to _____(increase, decrease).

19. When the pH of the blood falls to 7.1 or 7.2 the condition is known as _____. If the origin of this condition is the respiratory system it is called _____. If the origin is other than the respiratory system, it is called_____.

20. When the pH of the blood rises above normal the condition is known as _____. The body's means for returning the blood pH to normal include_____ and _____.

21. Paper bag rebreathing is one method of treating _____.

22. Possible causes of metabolic acidosis include _____, _____, and _____.

23. During respiratory acidosis, blood pH is lower than normal and the partial pressure of carbon dioxide is _____ than normal.

21.7 Formation of Blood Clots

Clot formation involves a complex series of reactions. In a simplified version, it can be viewed as a series of steps. When a blood vessel is cut or ruptures the blood platelets release prothrombin. Prothrombin is the inactive form of the enzyme thrombin. Prothrombin is then converted to thrombin by the action of thromboplastins, calcium ion and other factors. The enzyme thrombin then catalyzes the conversion of the globular plasma protein fibrin to the insoluble fibrin, which is the clot.

Blood can be prevented from clotting by removing calcium ions. Because citrate and EDTA form complexes with calcium, they can be added to blood samples to prevent clotting. Heparin is used to prevent clotting in the capillary tubes used to collect small samples of blood. Heparin and Coumadin can be administered to patients to prevent clotting.

Important terms

blood platelets	platelets factors	thromboplastin
Prothrombin	thrombin	fibrinogen
fibrin		

Check Your Understanding

For questions 24 to 31 select the best answer or answers.

24. Which of the following is not one of the formed elements of the blood?
  (a) leukocytes   (c) erythrocytes
  (b) platelets   (d) fibrinogen

25. When blood is centrifuged, the clear liquid that separates is
  (a) plasma   (c) serum
  (b) water   (d) fibrin

26 The plasma protein that participates in clot formation is
  (a) pepsinogen   (c) albumin
  (b) fibrinogen   (d) gamma globulin

27. Retention of excess fluid by the tissues is called
  (a) dehydration   (c) edema
  (b) hydrolysis   (d) hydration

28. Removal of this ion from blood will prevent clot formation.
  (a) Ca^{2+} (c) Na^+
  (b) SO_4^{2-} (d) K^+

29. Normally, the pH of the blood is about
  (a) 7.0   (c) 6.0
  (b) 6.8   (d) 7.4

30. The following is found in blood plasma, but not in blood serum.
  (a) albumin   (c) water
  (b) fibrinogen   (d) sulfate

31. This protein helps fight infectious diseases is
  (a) albumin   (c) gamma globulin
  (b) fibrinogen   (d) hemoglobin

21.8   Urine Specific Gravity

The specific gravity of a urine sample is a measure of the concentration of dissolved solids in the sample. The specific gravity of urine is normally between 1.005 and 1.035. Because the specific gravity of pure water is 1, low urine specific gravity indicates a low concentration of dissolved materials. Normally, as urine volume increases specific gravity decreases and vice versa. Diabetes mellitus however, is characterized by polyuria and high urine specific gravity. Diabetes insipidus is characterized by polyuria and low urine specific gravity.

Important Terms

specific gravity   polyuria   diuretics
antidiuretics   diabetes mellitus   diabetes insipidus
oliguria   anuria

21.9   Urine pH

Although the pH of the urine is different from the pH of the blood, changes in blood pH are reflected by changes in urine pH. The pH of urine ranges from 4.8 to 7.5, but is usually about 6.0. Low urine pH is an indication of excess acid in the system, perhaps due to the presence of the ketone bodies in the blood (ketonuria). High urine pH might be due to the "alkaline tide" that occurs after meals high in fruits and vegetables.

Important Terms

ketone bodies   alkaline tide

Check Your Understanding ————————————————————————————

For questions 32 to 36 fill in the blanks.
32.   The minimum volume of urine needed to remove body wastes from an adult is about _____mL per day.
33.   The production of large amounts of urine is known as _____. Two possible causes of this condition are: _____ and _____.
34.   The disease _____ is characterized by an extremely high urine flow. It is caused by a lack of _____ that are compounds that regulate urine flow.
35.   When the body loses excessive amounts of fluid through diarrhea, the volume of urine _____(increases, decreases). The term _____ is used to describe this condition.
36.   The term _____means "to pass through" and the term _____ means "sweet."

For questions 37 to 43 answer true of false.

37.   The presence of ketone bodies in the urine will raise urine pH.

38.   As blood pH falls, urine pH increases.

39.   Urine is slightly basic to keep metal ions in solution.

40.   Oliguria is usually accompanied by high urine specific gravity.

41.   Normal urine has a specific gravity slightly less than that of pure water.

42.   Urine that is less concentrated has a lower specific gravity.

43.   The "alkaline tide" describes the high urine pH following a meal high in fruits and
       vegetables.

21.10   Normal Components of Urine

Inorganic Components

▸   sodium ion   Varies with the amount of salt in the diet.

▸   potassium ion   Varies with the amount of potassium in the diet.

▸   chloride ion   Also varies with the amount of salt in the diet.

▸   ammonium   Produced by the deamination of proteins. The amount in the
                       urine varies with the amount of protein in the diet. A high
                       concentration of ammonium ion in the urine might also be
                       caused by acidosis.

▸   phosphate   Found in the urine as both HPO_4^{2-} and $H_2PO_4^-$. The total
                       amount of phosphate in the urine reflects the amount of
                       phosphate in the diet. The ratio $HPO_4^{2-}/H_2PO_4^-$
                       varies with blood pH.

▸   sulfur   Usually found in the urine as the sulfate ion, but some is
                       present as organic sulfur-containing compounds. Varies with
                       the amount of protein in the diet.

Organic Components

▸   urea   The principle end product of nitrogen metabolism in humans.
                       It makes up about half of the dissolved material in urine.

▸   uric acid   Produced during the metabolism of the purines in nucleic
                       acids.

▸   creatinine   Produced by the breakdown of creatine in muscle tissue. The
                       amount of creatinine in the urine per day per kilogram of body
                       weight (creatinine coefficient) is a measure of the percent of
                       muscle tissue in the body.

Important Terms

urea	uric acid	creatine
creatinine	creatinine coefficient	

21.11 Abnormal Components of Urine

Abnormal components of urine include:

▸ glucose   May be due to diabetes mellitus, renal diabetes, emotional stress, consumption of large amounts of carbohydrates over a brief period, or moderate kidney damage.

▸ plasma proteins   May be due to kidney disease, severe heart disease or certain infectious diseases.

▸ ketone bodies   Possible causes include: starvation, diabetes mellitus, severe liver damage, and a ketogenic diet

Important terms

urochrome	glycosuria	renal diabetes
false glycosuria	plasma proteins	ketonuria

Check Your Understanding ─────────────────────────────

For questions 44 to 50 select the correct answer.

44. An abnormal component of urine is
   (a) uric acid   (c) albumin
   (b) urea   (d) creatinine

45. The term used for the condition in which no urine is formed:
   (a) oliguria   (c) anuria
   (b) polyuria   (d) hydrolysis

46. The presence of ketones in the urine is known as
   (a) ketosis   (c) ketonuria
   (b) ketonemia   (d) ketone bodies

47. The deposit of uric acid and its salts in the joints is known as
   (a) uremia   (c) gout
   (b) acidemia   (d) arthritis

48. High urine production due to a disease such as diabetes insipidus:
   (a) oliguria   (c) anuria
   (b) polyuria   (d) hydrolysis

49. The major organic component of urine is
   (a) uric acid   (c) creatine
   (b) creatinine   (d) urea

50. Concentrated nitric acid is poured down the side of a tube containing urine. A white ring appears. This is a positive test for
   (a) glucose   (c) ketone bodies
   (b) albumin   (d) lactose

Answers to Check Your Understanding Questions in Chapter 21

1. diffusion **2.** open, open **3.** closed **4.** extracellular, lymph, digestive juices, synovial fluids, interstitial fluids, blood urine **5.** 45%, formed elements, hematocrit **6.** albumins, globulins, fibrinogens **7.** 92%, 7% **8.** albumins **9.** albumins **10.** oxygen, carbon dioxide **11.** low **12.** oxyhemoglobin **13.** 10%, hemoglobin, 70%, bicarbonate **14.** high, low **15.** increase **16.** a differentially permeable membrane **17.** albumins, osmotic **18.** decrease **19.** acidosis, respiratory acidosis, metabolic acidosis **20.** alkalosis, decrease expulsion of carbon dioxide by the lungs, increase excretion of bicarbonate by the kidneys **21.** respiratory alkalosis **22.** kidney failure, shock, diabetes mellitus **23.** higher **24.** d **25.** c **26.** b **27.** c **28.** a **29.** d **30.** b **31.** c **32.** 500 mL **33.** polyuria, high fluid intake, diabetes insipidus, diabetes mellitus **34.** diabetes insipidus, antidiuretics **35.** decreases, oliguria **36.** diabetes, mellitus **37.** F **38.** F **39.** F **40.** T **41.** F **42.** T **43.** T **44.** c **45.** c **46.** c **47.** c **48.** b **49.** d **50.** b

Appendix   ANSWERS TO THE EVEN-NUMBERED END-OF-CHAPTER PROBLEMS

Chapter 1

2.   Her weight will be 20 lbs (one-sixth of what it is on earth) but her mass will remain unchanged.

4.   (a) Two examples of an element. (from the Periodic Table)
      (b) Two examples of a compound.(for example: sodium chloride and carbon dioxide)

6.   (a) Two examples of foods that are homogeneous. (apple juice and mayonaise)
      (b) Two examples of foods that are heterogeneous. (chicken noodle soup and
          chocolate chip cookies)

8.   A physical change involves only a change of form whereas a chemical change involves changes in the chemical composition of the substances.

10.   Doctors collect data such as from blood tests and CAT scans, they make a hypothesis (diagnosis) and they test their hypothesis make prescribing treatment. If the treatment is not effective, they make a new hypothesis and prescribe a different treatment.

12.   Accuracy is a measure of how close a measurement is to the true value. Precision is a measure of reproducibility, how close a series of measurements are to each other.

14.   0.1 cm

16.   (a) $\dfrac{1\ kg}{1000\ g}$ (b) $\dfrac{946\ mL}{1\ qt}$ (c) $\dfrac{1\ ton}{2000\ lb}$

18.   (a)   0.405 kg   (h)   22,400 mL   (o)   0.75 liter
      (b)   1.563 km   (i)   30 cm   (p)   0.005 g
      (c)   160 mm   (j)   0.67 mm   (q)   3.6 dL
      (d)   1500 g   (k)   0.125 g   (r)   46,000 m
      (e)   0.015 liter   (l)   1200 cm   (s)   950 g
      (f)   0.127 m   (m)   0.456 mg   (t)   20 mL
      (g)   70 mg   (n)   1300 mm   (u)   3.61 m

20.   (a) 5 g   (b) 90 kg   (c) 400 mg   (d) 250 mL
          (e) 5 mL   (f) 1 L   (g) 15 cm   (h) 92 m

22.   (a) 1883 K, 2930°F   (b) 351.6 K, 173°F   (c) 212.4 K, -77.3°F

24.   (a) 13.6 g/mL or 13.6 g/cc   (b) 3.40 kg

26.   25.65 cm and 256.5 mm   28.   13 yd^2, 11 m^2

30.   1.5 cc   32.   82,000 L

34.   8.9 x 10^6 kg

36.   (a) -196°C   (b) -320°F

38.   (a) 6422°F   (b) 3823 K

40.   We know that gasoline has a lower specific gravity than water because gaoline floats on water.

42.   1000mg/ day   44.   1.76 x 10^{-7} mg/kg

46.   125 cc/hr   48.   57 L

Chapter 2

2.   Examples from page 36

4.   The temperature decreases.

6.   (a)   potential energy and kinetic energy are low
        (b)   potential energy is increasing, the movement from the lift gives her kinetic energy
        (c)   potential energy is high, no kinetic energy
        (d)   potential energy is decreasing as kinetic energy is increasing

8.   (a) 175 cal, 0.175 kcal   (c) 51,000 cal, 51.0 kcal
        (b) 400 cal, 0.400 kcal   (d) 11,000 cal, 11.0 kcal

10.   7 kcal/g   12.   endothermic

14. (a) short waves, microwaves, yellow light, ultraviolet light, X rays

(b) X rays, ultraviolet light, yellow light, microwaves, short waves

16. We know from the First Law of Thermodynamics that energy is conserved. If we take in more energy than we need the energy does not disappear because that would be a violation of the First Law. Instead, the energy is stored, mainly as fat.

18. The energy source is the food we eat. The source of the energy in this food is the sun.

20. 2.2 hours

$$150 \text{ kcal} \times \frac{\text{hr} - \text{kg}}{1 \text{ kcal}} \times \frac{1}{150 \text{ lb}} \times \frac{2.2 \text{ lb}}{1 \text{ kg}} = 2.2 \text{ hours}$$

22. 0.06 cal/g °C

24. 3.1×10^5 cal

26. 13.5 tons

28. 3.0×10^6 Btu/year

30. UV-B is more likely to cause skin cancer because it has more energy than UV-A and can do more damage to skin cells.

Chapter 3

2. 4 kcal

$$50 \text{ g} \times \frac{80 \text{ cal}}{1 \text{ g}} \times \frac{1 \text{ kcal}}{1000 \text{ cal}}$$

4.

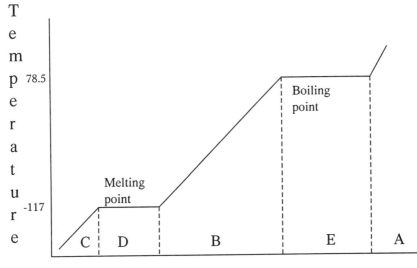

6.   135 kcal (calculator answer 134.750)

8.   At 110°C water molecules are in the gaseous state. They have high kinetic energy
        and are moving very fast in all directions. As the molecules are cooled to 100°C,
        they move slower and slower. At 100°C the attractions between molecules are strong
        enough to hold the molecules together when they collide, and a liquid begins to form.
        When all the molecules are in the liquid state, the temperature will decrease and the
        molecules will move slower and slower. At 0°C the kinetic energy of the molecules
        is so low and the attraction between the molecules so strong that the molecules
        remain in fixed positions in the structure of ice. As the temperature continues to
        decrease, the vibrations of the molecules in the solid ice become less energetic.

10.   (a) 3800 mm Hg   (d) 2100 torr   (g) 14.3 psi
        (b) 0.25 atm   (e) 8400 Pa   (h) 1.17 atm
        (c) 70 mm Hg   (f) 91 kPa

12.   845 mm Hg   14.   840 mm Hg

16.   -8° C   18.   1.2 liters

20.   statement of Henry's law

                solubility = kP, when the temperature is constant

        When you open a container of carbonated beverage, the pressure is reduced.
        Therefore, the solubility of the carbon dioxide in the beverage decreases and the
        escaping gas makes a hissing sound.

22.   739 torr

24.   Oxygen tension in arterial blood is higher than in venous blood.

26.   When the volume of a gas decreases, the gas particles hit the walls of the container
        more often causing an increase in pressure.

28.   The sunlight warms the puddle faster than if the weather remained cloudy. The
        molecules of the warmer water have more kinetic energy and molecules near the
        surface can break away and escape from the liquid and evaporate.

30.   When the door to the room is opened, the air in the room, which is at a higher

pressure, will flow out of the room preventing exterior air, dust, and microorganisms from entering.

32.   1100 mL   34.   38 psi

36.   (a) 1100 torr   (b) 1.46 atm

Chapter 4

2.   The atom has a small dense nucleus with the electrons found in the region surrounding the nucleus, but the atom is mostly empty space. If the atom were a baseball stadium, the nucleus would be a flea at second base and the nearest electron would be in the upper deck of stands.

4.   (a) Z = 17, M = 35, ^{35} Cl ,$^{35}_{17}$ Cl, chlorine-35
        (b) Z = 27, M = 60, ^{60} Co, $^{60}_{27}$ Co, cobalt-60
        (c) Z = 1, M = 3, ^{3} H, $^{3}_{1}$ H, hydrogen-3

6.   (a) Z = 8, M = 16, $^{16}_{8}$ O, ^{16}O, oxygen-16
        (b) Z = 90, M = 232, $^{232}_{90}$ Th, ^{232}Th, thorium-232
        (c) Z = 47, M = 107, $^{107}_{47}$ Ag, ^{107}Ag, silver-107

8.   zinc-64 p = 30   n = 34   e = 30
        zinc-66 p = 30   n = 36   e = 30
        zinc-67 p = 30   n = 37   e = 30
        zinc-68 p = 30   n = 38   e = 30
        zinc-70 p = 30   n = 40   e = 30

10.   all three have 33 neutrons

12.   copper = 63.6 amu, magnesium = 24.3 amu

14.   4th energy level, 32 electrons; 6th energy level, 72 electrons

16.   (a) Be $1s^2\, 2s^2$
        (b) I $1s^2\, 2s^2\, 2p^6\, 3s^2\, 3p^6\, 4s^2\, 3d^{10}\, 4p^6\, 5s^2\, 4d^{10}\, 5p^5$
        (c) Si $1s^2\, 2s^2\, 2p^6\, 3s^2\, 3p^2$
        (d) S $1s^2\, 2s^2\, 2p^6\, 3s^2\, 3p^4$
        (e) Se $1s^2\, 2s^2\, 2p^6\, 3s^2\, 3p^6\, 4s^2\, 3d^{10}\, 4p^4$
        (f) Cu $1s^2\, 2s^2\, 2p^6\, 3s^2\, 3p^6\, 4s^1\, 3d^{10}$

(g) Mn $1s^2 2s^2 2p^6 3s^2 3p^6 4s^2 3d^5$

(h) Ca $1s^2 2s^2 2p^6 3s^2 3p^6 4s^2$

(i) Ra $1s^2 2s^2 2p^6 3s^2 3p^6 4s^2 3d^{10} 4p^6 5s^2 4d^{10} 5p^6 6s^2 4f^{14} 5d^{10} 6p^6 7s^2$

(j) K $1s^2 2s^2 2p^6 3s^2 3p^6 4s^1$

18. $1s^2 2s^2 2p^6 3s^2 3p^4$

20. No, the electron in the 4th energy level is in a higher-than-normal energy level. It would be in the 3rd energy level in the ground state.

22. atomic size, ionization energy, electron affinity

24.
(a)	representative	(d)	representative	(g)	representative
(b)	representative	(e)	transition	(h)	representative
(c)	transition	(f)	representative		

26. Atoms A and C: they have the same number of electrons in the outermost energy level.

28. (a) Si, P, S, Cl   (b) Rb, Pd, Ag, Sn

30. (a) Ca, Mg, Be   (b) Te, I, Xe   (c) S, Cl, F

32. (a) Cl   (b) Cl   (c) Cl

34. Trace elements are those found in only very small amounts in living organisms.

36. group VIA

38. P = 30.94 $30.97 \times \dfrac{12.00}{12.01} = 30.94$

Se = 78.89 $78.96 \times \dfrac{12.00}{12.01} = 78.89$

40. 128 Notice that the increase in maximum number of electrons between the 3rd and 4th level is 14; between the 4th and 5th level it is 18; between the 5th and 6th level it is 22 and between the 6th and 7th level it is 26. The next increase should be 30. $98 + 30 = 128$.

Chapter 5

2. They all have eight electrons in their outermost energy level.

4. (a) Calcium has two electrons in its outermost energy level. In order to achieve an octet it must lose two electrons.

    (b) Potassium has one electron in its outermost energy level. When it losses on electron it has achieved an octet.

6.   Na $^+$: F̈ : $^-$ Mg $^{2+}$: Ö : $^{2-}$

    Na$^+$ has ten electrons, F$^-$ has ten electrons, Mg^{2+} has ten electrons and so does O^{2-}.

8. Covalent compounds are composed of separate, neutral units called molecules, and ionic compounds are composed of neutral aggregates of ions.

10. (a) H: C̈:: C̈: H   H—C=C—H
         H   H | |
                                    H   H

    (b) : Ö: : C : Ö: H   O=C—O—H
             Ḧ |
                                    H

    (c) H : C ::: C : H   H—C≡C—H

    (d) : Ö: : Si: : Ö: O=Si=O

12. (a)   H—N—H   (d)
              |
              H   Br—C≡C—Br

          Cl
          |
    (b) Cl—C—Cl   (e)   F—S—F
          |
          Cl

    (c)   H—Cl   (f)   I—I

267

14. (a) nonpolar   (d) polar, Cl   (g) polar, F
     (b) polar, Br   (e) polar, O   (h) nonpolar
     (c) polar, N   (f) polar, O

16.

	More positive	More negative
(a)	Mg	O
(b)	N	O
(c)	O	F
(d)	H	N
(e)	H	C
(f)	I	Br
(g)	H	Br
(h)	C	Cl
(i)	Pb	S

18. (a) polar   (c) polar   (e) nonpolar
     (b) nonpolar   (d) polar   (f) nonpolar

20. (a) ionic   (d) polar covalent   (g) polar covalent
     (b) polar covalent   (e) nonpolar covalent   (h) nonpolar covalent
     (c) polar covalent   (f) polar covalent   (i) ionic

22. Compounds a, d and e exhibit hydrogen bonding with another molecule of the same compound.

24. (a) $LiC_2H_3O_2$ (f) $Sn_3(PO_4)_2$
     (b) $Mg(HCO_3)_2$ (g) Ag_2SO_3
     (c) $Al_2(CrO_4)_3$ (h) Mn_2O_3
     (d) Sr_3P_2 (i) $Pb(H_2PO_4)_4$
     (e) $Co_2(C_2O_4)_3$ (j) $(NH_4)_3N$

26. (a) N = 3,   H = 12,   P = 1,   O = 4
     (b) Al = 2,   H = 3,   P = 3,   O = 12
     (c) Ca = 1,   C = 4,   H = 6,   O = 4
     (d) Fe = 7,   C = 18,   N = 18

28. Set 1: (a) LiF   (e) FeS   (i) Hg_3N_2
          (b) K_2S (f) BaI_2 (j) $PbCl_4$
          (c) $MgBr_2$ (g) Al_2O_3 (k) CsI
          (d) AgCl   (h) Cu_2O (l) GaN

Set 2: (a) $(NH_4)_2CO_3$ (e) $Ba(NO_2)_2$ (i) $SrCr_2O_7$
        (b) $Ca(HSO_4)_2$ (f) $Mg(H_2PO_4)_2$ (j) $Be(NO_3)_2$
        (c) $Mg(HCO_3)_2$ (g) $Fe_2(SO_4)_3$ (k) $Cr_2(SO_3)_3$
        (d) Li_3PO_4 (h) KCN   (l) $Ni(C_2H_3O_2)_2$

Set 3: (a) IF_5 (e) BCl_3 (i) Si_2Br_6
        (b) H_2Se (f) SO_2 (j) PF_3
        (c) N_2O_4 (g) CS_2 (k) OF_2
        (d) HBR   (h) SCl_2 (l) SeO_3

30. (a) lithium acetate   (f) tin(II) phosphate
     (b) magnesium bicarbonate   (g) silver sulfite
     (c) aluminum chromate   (h) manganese(III) oxide
     (d) strontium phosphide   (i) lead(IV) dihydrogen phosphate
     (e) cobalt (III) oxalate   (j) ammonium nitride

32. 1-Butanol has a hydrogen on an oxygen and can form hydrogen bonds. Ethyl ether does not have the ability to form hydrogen bonds.

34.

H : Ö : Ö : H   H—O—O—H

The fact that the compound is polar tells us that it cannot be linear, it must be bent.

36.
$$F : \ddot{S} : F$$
with F above and F below

Chapter 6

2.   (a) $2Na + 2H_2O \longrightarrow 2NaOH + H_2$

     (b) $2KClO_3 \longrightarrow 2KCl + 3O_2$

     (c) $MnO_2 + 4HCl \longrightarrow Cl_2 + MnCl_2 + 2H_2O$

     (d) $C_3H_8 + 5O_2 \longrightarrow 3CO_2 + 4H_2O$

     (e) $4NH_3 + 5O_2 \longrightarrow 4NO + 6H_2O$

     (f) $CO_3^{2-} + 2H^+ \longrightarrow CO_2 + H_2O$

4. (a) $2AgNO_3 + CaCl_2 \longrightarrow Ca(NO_3)_2 + 2AgCl$

(b) $Fe_2O_3 + 3H_2 \longrightarrow 2Fe + 3H_2O$

(c) $2Al(NO_3)_3 + 3H_2SO_4 \longrightarrow Al_2(SO_4)_3 + 6HNO_3$

(d) $C_3H_8O + 5O_2 \longrightarrow 3CO_2 + 4H_2O$

(e) $2Al + 3H_2SO_4 \longrightarrow Al_2(SO_4)_3 + 3H_2$

(f) $2Fe_2O_3 + 3C \longrightarrow 4Fe + 3CO_2$

6.

	(a)	(b)
oxidizing agent	Cl_2	O_2
reducing agent	Ca	Mg

8. (a) 0.23 mol Br   (c) 1.60 mol O
   (b) 3.15 mol Cu   (d) 0.251 mol C

10.

	Group I	Group II	Group III
(a)	233 amu	84.0 amu	85.0 amu
(b)	80.9 amu	158 amu	170 amu
(c)	67.8 amu	63.0 amu	62.0 amu
(d)	18.0 amu	160 amu	100 amu

12. $\quad$ 3.920 L $\quad$ 1.054 x 10^{23} ~~molecules~~ x $\dfrac{\text{mol}}{6.023 \times 10^{23} \text{ \sout{molecules}}}$ x $\dfrac{22.4 \text{ L}}{\text{mol}}$

14. (a) 1) 126 mL   2) 369 mL   3) 1.05 L
    (b) 1) 0.322 atm   2) 1.13 atm   3) 1.85 atm
    (c) 1) 288°C (561 K) 2) 850°C (1123 K) 3) 852°C (1125 K)

16.

	Group I	Group II	Group III
(a)	0.000300 mol	1.20×10^3 mol	5.49 mol
(b)	0.300 mol	0.0749 mol	2.5×10^{-8} mol
(c)	38.8 mol	4.98×10^{-7} mol	9.30 mol
(d)	0.0050 mol	4.50 mol	6.00×10^{-5} mol
(e)	0.0100 mol	8.00×10^{-4} mol	0.760 mol
(f)	6.51×10^{-4} mol	3.16 mol	0.0040 mol
(g)	1.81 mol	1.50×10^{-6} mol	1.30×10^{-3} mol
(h)	3.60×10^{-6} mol	1.99×10^{-2} mol	12.5 mol

18.   (a) 1.1×10^{25} atoms

     (b) 3.6×10^{23} atoms

$$18\ \cancel{g} \times \frac{1\ \cancel{mol}}{180\ \cancel{g}} \times \frac{6.03 \times 10^{23}\ atoms}{1\ \cancel{mol}}$$

     (c) 2.2×10^{22} atoms

$$1.0\ \cancel{g} \times \frac{1\ \cancel{mol}}{250\ \cancel{g}} \times \frac{6.03 \times 10^{23}\ atoms}{1\ \cancel{mol}}$$

20.   (a) $2.5\ \cancel{mol\ Fe_2O_3} \times \dfrac{2\ mol\ Al}{1\ \cancel{mol\ Fe_2O_3}} = 5.0\ mol\ Al$

     (b) $2.5\ \cancel{mol\ Fe_2O_3} \times \dfrac{1\ \cancel{mol\ Al_2O_3}}{1\ \cancel{mol\ Fe_2O_3}} \times \dfrac{102\ g}{\cancel{mol\ Al_2O_3}} = 255\ g\ Al_2O_3$

     (c) $13.5\ \cancel{g}\ Al \times \dfrac{\cancel{mol\ Al}}{27.0\ \cancel{g}} \times \dfrac{1\ \cancel{mol\ Fe_2O_3}}{2\ \cancel{mol\ Al}} \times \dfrac{159.8\ g}{\cancel{mol\ Fe_2O_3}} = 40.0\ g\ Fe_2O_3$

22.   (a) $150\ \cancel{mg}\ H_2 \times \dfrac{1\ \cancel{g}}{1000\ \cancel{mg}} \times \dfrac{1\ \cancel{mol\ H_2}}{2.0\ \cancel{g}} \times \dfrac{1\ \cancel{mol\ Fe_2O_3}}{3\ \cancel{mol\ H_2}} \times \dfrac{160\ g}{\cancel{mol\ Fe_2O_3}} = 4.00\ g$

     (b) $2.0 \times 10^{-2}\ \cancel{mol\ Fe_2O_3} \times \dfrac{2\ \cancel{mol\ Fe}}{1\ \cancel{mol\ Fe_2O_3}} \times \dfrac{55.9\ g}{\cancel{mol\ Fe}} = 2.24\ g\ Fe$

     (c) $160\ \cancel{mg\ Fe_2O_3} \times \dfrac{1\ \cancel{g}}{1000\ \cancel{mg}} \times \dfrac{\cancel{mol\ Fe_2O_3}}{159.8\ \cancel{g}} \times \dfrac{3\ mol\ H_2O}{1\ \cancel{mol\ Fe_2O_3}} = 3.00 \times 10^{-3}\ mol$

24.   (a) $0.30\ mol\ Mg(OH)_2$ $0.10\ \cancel{mol\ Mg_3N_2} \times \dfrac{3\ mol\ Mg(OH)_2}{1\ \cancel{mol\ Mg_3N_2}}$

     (b) $14.9\ mol\ Mg(OH)_2$ $500\ \cancel{g} \times \dfrac{1\ \cancel{mol\ Mg_3N_2}}{101\ \cancel{g}} \times \dfrac{3\ mol\ Mg(OH)_2}{1\ \cancel{mol\ Mg_3N_2}}$

         $9.9\ mol\ NH_3$ $500\ \cancel{g} \times \dfrac{1\ \cancel{mol\ Mg_3N_2}}{101\ \cancel{g}} \times \dfrac{2\ mol\ NH_3}{1\ \cancel{mol\ Mg_3N_2}}$

     (c) $2.0\ g\ Mg_3N_2$ $0.060\ \cancel{mol\ Mg(OH)_2} \times \dfrac{1\ \cancel{mol\ Mg_3N_2}}{3\ \cancel{mol\ Mg(OH)_2}} \times \dfrac{101\ g}{1\ \cancel{mol\ Mg_3N_2}}$

         $2.2\ g\ H_2O$ $0.060\ \cancel{mol\ Mg(OH)_2} \times \dfrac{6\ mol\ H_2O}{3\ \cancel{mol\ Mg(OH)_2}} \times \dfrac{18.0\ g}{1\ \cancel{mol\ H_2O}}$

(d)   30.1 g Mg(OH)$_2$

$$52.2 \text{ g Mg(OH)}_2 \times \frac{1 \text{ mol Mg(OH)}_2}{58.3 \text{ g Mg(OH)}_2} \times \frac{1 \text{ mol Mg}_3\text{N}_2}{3 \text{ mol Mg(OH)}_2} \times \frac{101 \text{ g}}{1 \text{ mol Mg}_3\text{N}_2}$$

(e)   17.3 g Mg(OH)$_2$

$$10.0 \text{ g Mg}_3\text{N}_2 \times \frac{1 \text{ mol Mg}_3\text{N}_2}{101 \text{ g}} = 0.0990 \text{ mol Mg}_3\text{N}_2$$

$$0.0990 \text{ mol Mg}_3\text{N}_2 \times \frac{6 \text{ mol H}_2\text{O}}{1 \text{ mol Mg}_3\text{N}_2} = 0.594 \text{ mol H}_2\text{O}$$

$$14.4 \text{ g H}_2\text{O} \times \frac{1 \text{ mol H}_2\text{O}}{18.0 \text{ g}} = 0.800 \text{ mol H}_2\text{O}$$

$$0.0990 \text{ mol Mg}_3\text{N}_2 \times \frac{3 \text{ mol Mg(OH)}_2}{1 \text{ mol Mg}_3\text{N}_2} \times \frac{58.3 \text{ g}}{1 \text{ mol Mg(OH)}_2} = 17.3 \text{ g}$$

26.   0.25 g salicylic acid

$$0.33 \text{ g C}_9\text{H}_8\text{O}_4 \times \frac{1 \text{ mol C}_9\text{H}_8\text{O}_4}{180 \text{ g}} \times \frac{1 \text{ mol C}_7\text{H}_6\text{O}_3}{1 \text{ mol c}_9\text{H}_8\text{O}_4} \times \frac{138 \text{ g}}{1 \text{ mol C}_7\text{H}_6\text{O}_3}$$

28.   225 g H$_2$O $CH_4 + 2O_2 \longrightarrow 2H_2O + CO_2$

$$100 \text{ g CH}_4 \times \frac{1 \text{ mol CH}_4}{16.0 \text{ g CH}_4} \times \frac{2 \text{ mol H}_2\text{O}}{1 \text{ mol CH}_4} \times \frac{18.0 \text{ g}}{1 \text{ mol H}_2\text{O}}$$

30.   (a) $2Al_2O_3 + 3C \longrightarrow 4Al + 3CO_2$

(b)   10.2 metric tons Al$_2$O$_3$

$$5.40 \text{ tons Al} \times \frac{1 \times 10^3 \text{ kg}}{1 \text{ ton Al}} \times \frac{1 \text{ mol Al}}{0.027 \text{ kg}} \times \frac{2 \text{ mol Al}_2\text{O}_3}{4 \text{ mol Al}} \times$$

$$\frac{0.102 \text{ kg Al}_2\text{O}_3}{1 \text{ mol Al}_2\text{O}_3} \times \frac{1 \text{ ton}}{1 \times 10^3 \text{ kg}} = 10.2 \text{ tons}$$

(c)   0.33 metric ton C

32.   (a) limiting reactant :O_2 4 mol C_4H_{10} × $\dfrac{13 \text{ mol } O_2}{2 \text{ mol } C_4H_{10}}$ = 26 mol O_2

2.5 mol CO_2

$$4 \text{ mol } O_2 × \dfrac{8 \text{ mol } CO_2}{13 \text{ mol } O_2} = 2.5 \text{ mol } CO_2$$

(b) limiting reactant:O_2 120.0 g × $\dfrac{1 \text{ mol } C_4H_{10}}{58.0 \text{ } g}$ = 2.07 mol

102 g CO_2

$$120.0 \text{ L}g × \dfrac{1 \text{ mol } O_2}{32.0 \text{ } g} = 3.75 \text{ mol } O_2$$

$$3.75 \text{ mol } O_2 × \dfrac{8 \text{ mol } CO_2}{13 \text{ mol } O_2} × \dfrac{44.0 \text{ g}}{1 \text{ mol } CO_2} = 102 \text{ g } CO_2$$

(c) 0 g O_2, 43.0 g C_4H_{10}, 21.1 g CO_2, 10.8 g H_2O

O_2 is the limiting reagent

$$25.0 \text{ } g × \dfrac{1 \text{ mol } O_2}{32.0 \text{ } g} = 0.781 \text{ mol } O_2$$

$$50.0 \text{ } g × \dfrac{1 \text{ mol } C_4H_{10}}{58.0 \text{ } g} = 0.862 \text{ mol } C_4H_{10}$$

$$0.781 \text{ mol } O_2 × \dfrac{2 \text{ mol } C_4H_{10}}{13 \text{ mol } O_2} = 0.120 \text{ mol } C_4H_{10}$$

$$(0.862 \text{ mol} - 0.120 \text{ mol}) × \dfrac{58.0 \text{ g}}{1 \text{ mol } C_4H_{10}} = 43.0 \text{ g } C_4H_{10}$$

$$0.781 \text{ mol } O_2 × \dfrac{8 \text{ mol } CO_2}{13 \text{ mol } O_2} × \dfrac{44.0 \text{ g}}{1 \text{ mol } CO_2} = 21.1 \text{ g } CO_2$$

$$0.781 \text{ mol } O_2 × \dfrac{10 \text{ mol } H_2O}{13 \text{ mol } O_2} × \dfrac{18.0 \text{ g}}{1 \text{ mol } H_2O} = 10.8 \text{ g } H_2O$$

Integrated Problems

2.   2.778 g/mL $15.65 \cancel{g} \times \dfrac{1 \text{ mL}}{0.7914 \cancel{g}} = 19.78 \text{ mL}; \quad \dfrac{14.50 \text{ g}}{5.22 \text{ mL}}$

4.   21.5°C   Heat lost = heat gained

$$200.0 \cancel{g} \times \dfrac{0.115 \cancel{cal}}{\cancel{g}°C} \times (85.0°C - T_f) = 1000 \cancel{g} \times \dfrac{1.00 \cancel{cal}}{\cancel{g} °C} \times (T_f - 20.0°C)$$

$$1955 - 23T_f = 1000T_f - 20,000$$

$$21.5°C = T_f$$

6.   3.5 days/lb $\dfrac{1 \text{ day}}{1000 \cancel{kcal}} \times \dfrac{7700 \cancel{kcal}}{1 \cancel{kg}} \times \dfrac{1 \cancel{kg}}{2.2 \text{ lb}}$

8.   (a) 136 lb $62,000 \cancel{g} \times \dfrac{1 \cancel{kg}}{1000 \cancel{g}} \times \dfrac{2.2 \text{ lb}}{1 \cancel{kg}}$

       (b) 4800 kcal/week $\dfrac{62,000 \cancel{g}}{1 \cancel{year}} \times \dfrac{1 \cancel{year}}{52 \text{ weeks}} \times \dfrac{4.0 \text{ kcal}}{1 \cancel{g}}$

10.   The potassium ion might interfere with the data since potassium and sodium are in the same chemical family and have similar chemical properties. To test this conclusion, set up a series of controlled experiments:

       Number 1 with uncontaminated medium as a control
       Number 2 with control medium plus a trace amount of K^+
       Number 3 with control medium plus a trace amount of Ba^{2+}
       Number 4 with control medium plus a trace amount of I^-

    Monitor each experiment carefully to determine whether trace amounts of these ions will alter the data obtained with the control.

12.   1220 g H_2SO_4 $SO_3 + H_2O \longrightarrow H_2SO_4$

$$1.00 \text{ kg } SO_3 \times \dfrac{1 \cancel{mol\ SO_3}}{0.0801 \text{ kg } SO_3} \times \dfrac{1 \cancel{mol\ H_2SO_4}}{1 \cancel{mol\ SO_3}} \times \dfrac{98.1 \text{ g}}{1 \cancel{mol\ H_2SO_4}}$$

Chapter 7

2. Alpha radiation consists of streams of positive helium nuclei which have very little penetrating power. Beta radiation consists of streams of negative electrons which have more penetrating power than alpha particles. Gamma radiation is high energy electromagnetic radiation, which has very high penetrating power.

4. $^{210}_{83}Bi \longrightarrow {}^{210}_{84}Po + {}^{0}_{-1}e$

6. $^{239}_{94}Pu \longrightarrow {}^{239}_{95}Am + {}^{0}_{-1}e + \gamma$

8. $^{241}_{93}Np \longrightarrow {}^{237}_{91}Pa + {}^{4}_{2}He \longrightarrow {}^{237}_{92}U + {}^{0}_{-1}e \longrightarrow$

   $^{237}_{93}Np + {}^{0}_{-1}e \longrightarrow {}^{233}_{91}Pa + {}^{4}_{2}He \longrightarrow {}^{229}_{89}Ac + {}^{4}_{2}He \longrightarrow$

   $^{225}_{87}Fr + {}^{4}_{2}He \longrightarrow {}^{221}_{85}At + {}^{4}_{2}He \longrightarrow {}^{217}_{83}Bi + {}^{4}_{2}He \longrightarrow$

   $^{217}_{84}Po + {}^{0}_{-1}e \longrightarrow {}^{217}_{85}At + {}^{0}_{-1}e \longrightarrow {}^{213}_{83}Bi + {}^{4}_{2}He \longrightarrow$

   $^{213}_{84}Po + {}^{0}_{-1}e \longrightarrow {}^{213}_{85}At + {}^{0}_{-1}e \longrightarrow {}^{209}_{83}Bi + {}^{4}_{2}He$

10. 3 g   After 5 half-lives, $1/2^5$ or 1/32 of the sample will remain.

12. 58 years $1.25/5.00 = 1/4$ or $1/2^2$ of the sample is left. Two half-lives is 58 years.

14. It is called a chain reaction because each fission results in two fragments plus two or three neutrons that can strike and ract with other uranium-235 nuclei.

16. 3200 years   It takes two half-lives to reduce the sample to one-quarter of its original mass. lf-lives is 3200 years for this isotope.

18. $^{235}_{92}U + {}^{1}_{0}n \longrightarrow {}^{94}_{38}Sr + 2{}^{1}_{0}n + {}^{140}_{54}Xe$

20. Ionizing radiation can damage living cells by destroying DNA which causes the cell to be unable to divide and it dies. If the DNA is only damaged, it can result in the formation of mutant cells.

22. Living cells can repair DNA as it becomes changed. Cells also have antiocxidnats which quickly neutralize any free radiacsl or peroxides that are formed.

24. $$\frac{\text{trans}}{\text{minute}} = 5.0 \; \cancel{g} \; \text{x} \; \frac{100 \; \cancel{mCi}}{\cancel{g}} \; \text{x} \; \frac{1 \; \text{Ci}}{1000 \; \cancel{mCi}} \; \text{x} \; \frac{3.7 \; \text{X} \; 10^{10} \; \text{trans}}{1 \; \text{Ci - sec}} \; \text{x} \; \frac{60 \; \text{sec}}{\cancel{min}}$$

$= 1.1 \; \text{X} \; 10^{12}$ transformations / minute

26. It is not advisable that this patient have routine dental X rays, because the effects of radiation are cumulative and his exposure way above normal.

28. 1) Second alternative: the concrete dividing wall acts as shielding from the X rays.
    2) Third alternative: the farther from the source, the lower the intensity of the X rays.
    3) First alternative

30. PET scans overcome the inability of CT scans to detect the early stages of disease in tissues.

32. Iodine-131 can be administered and then a gamma scan of the thyroid can be run to determine uptake of the radioisotope by the thyroid.

34. Cancer cells are more vulnerable to damage from ionizing radiation because they are dividng more rapidly than the surrounding tissue.

36. about 45%. $T_{1/2} = 28$ years, so 50% would remain in 1991.

38. No, carbon dating is only accurate for objects that have died within the last 40,000 years.

40. 0.10 cc $10.0 \; \cancel{mCi} \; \text{x} \; \frac{1 \; \cancel{Ci}}{1000 \; \cancel{mCi}} \; \text{x} \; \frac{9.0 \; \text{cc}}{0.90 \; \cancel{Ci}}$

42. It would be preferable to use the cobalt with the shorter half-life, cobalt-57, to minimize the exposure of the patient to the radioactive compound.

Chapter 8

2. Most collisions do not have enough energy to reach to reach the activated complex.

4. (a) exothermic   (c) endothermic
    (b) endothermic   (d) endothermic

6. The conversion of graphite to diamond has an extremely high activation energy.

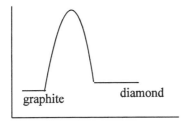

graphite   diamond

8.   The rate of a chemical reaction is greatly increased if a solid reactant is powdered, and there is a chance of an explosion when finely divided combustible material is being handled.

10.   The drugs deteriorate with time, and placing them in the refrigerator slows the rate of deterioration.

12.   The rates of the forward and reverse reactions are equal.

14.   A catalyst has no effect on the equilibrium concentrations of the reactants and products in an equilibrium system.

16.   Because it is an exothermic process, sodium hydroxide will dissolve more readily in cold water.

18.   1) increase the concentration of NO
    2) increase the concentration of Cl_2
    3) increase the pressure

20.   (a) increase   Increasing the temperature increases the rate of the reverse endothermic reaction.
    (b) decrease   Increasing the pressure will increase the forward reaction thereby lowering the number of molecules.
    (c) decrease   Decreasing the concentration of Cl_2 will increase the forward reaction to replace the lost products.
    (d) decrease   Increasing the concentration of the reactant HCl will increase the forward reaction.
    (e) no change   A catalyst increases the rate of both the forward and reverse reactions equally.

22.   (a)   the right   (b)   the left   (c)   the right
       (d)   the right   (e)   no change   (f)   no change

24.

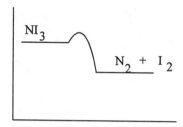

The activation energy for this reaction is so small that the energy from a feather can cause a few molecules to react, releasing enough energy to cause the rest of the NI_3 to react all at once.

26.   The rate of chemical reactions is temperature dependent: the lower the temperature, the slower the rate. When a snake's body is cold, it is sluggish because the reactions occurring in its body proceed slowly. The snake suns itself to increase its body temperature, and thus increase the rate of body cellular reactions.

28.   An increase in body temperature in a fever can be enough to kill invading microorganisms, but too high a fever will also be damaging to body cells.

30.   (a)   The reddish-brown color of NO_2 would decrease as the reverse reaction increases.
       (b)   The reddish-brown color will increase as the forward reaction increases.
       (c)   The reddish-brown color will increase as the forward reaction increases.

Integrated Problems

2.   (a) 1) $Z = 83$ $M = 209$ $p = 83$ $n = 126$
            2) $Z = 26$ $M = 58$ $p = 26$ $n = 32$

       (b) $^{209}_{83} Bi + ^{58}_{26} Fe \longrightarrow ^{267}_{109} E$

       (c) cobalt, rhodium, iridium

4.   (a) 3 men   From Table 7.4, only 50% of people exposed to more than 450 rem are expected to survive.

(b) 1.5×10^{13} transformations/sec

(c)   One would expect to see broken chromosomes or abnormally formed chromosomes.

6. Some coal contains naturally occurring radioactive substances which are released when the coal is burned.

Chapter 9

2. Carbon tetrachloride is a nonpolar molecule and ammonia is a polar molecule. Ammonia would be more soluble in water than carbon tetrachloride.

4. 1700 cal $22 \text{ g } \times \dfrac{1 \text{ cal}}{\text{g - }^\circ C} \times (90\text{-}15)^\circ C$

6. 490 cal $30 \text{ g } \times \dfrac{0.60 \text{ cal}}{\text{g - }^\circ C} \times (52\text{-}25)^\circ C$

8. (a)   It takes more energy to pull water molecules apart because of the hydrogen bonding and, as a result, the boiling point is high.
    (b)   Hydrogen bonding between water molecules results in the open lattice structure of ice, giving ice a low density.
    (c)   The strong attraction between water molecules makes it difficult for them to escape from the liquid and, as a result, the heat of vaporization is large.
    (d)   Polar molecular substances can form hydrogen bonds with water molecules and, as a result, are soluble in water.

10. An element is composed of only one type of atom. A compound contains different types of atoms but has one chemical identity. A mixture contains several elements or compounds each of which retains its chemical identity in the mixture.

12. Brownian movement in colloidal dispersions is caused by the bombardment of the colloid's particles by the surrounding solvent molecules.

14. The solvent is the dissolving medium. The solute is the substance being dissolved, and the solution is the mixture of the solute and the solvent.

16. Highly polar molecules like sugar dissolve in water of the attraction of oppositely charged polar regions on the solute molecule and water and because of hydrogen bonding, when possible.

18.   (b) and (c) form solutions in water that conduct electrical current.

20.   A strong electrolyte completely dissociates in water, but a weak electrolyte only partially dissociates.

22.   An increase in temperature increases the solubility of sodium chloride in water.

24.   (a) A precipitate of $PbSO_4$ will form.   (c) A precipitate of AgCl will form.
       (b) No precipitate will form.   (d) A precipitate of $PbCl_2$ will form.

26. $2H^+_{(aq)} + CO_3^{2-}_{(aq)} \longrightarrow H_2O_{(l)} + CO_{2(g)}$

28.   The Pacific Ocean plays a moderating role in the Los Angeles weather. Because of the ocean's large specific heat, it will absorb solar energy during the day and release it during the night, preventing large fluctuations in temperature.

30.   Sodium sulfate will cause the barium ions to precipitate as barium sulfate, thus removing the barium ions from the body.

$$Ba^{2+}_{(aq)} + SO_4^{2-}_{(aq)} \longrightarrow BaSO_{4(s)}$$

Chapter 10

2.   A saturated solution contains all of the solute particles the solvent can usually hold at a certain temperature. An unsaturated solution contains fewer particles than it can hold at a certain temperature. If the temperature decreases in a saturated solution but none of the solute particles crystalize, then the solution is supersaturated and contains more solvent particles than normal at that temperature.

4.   There will be no net change in the amount of dissolved solute but since an equilibrium exists, some of the added sugar will go into solution as other sugar molecules precipitate out of solution.

6.   (a)   Dissolve 12 g of HCl in enough water to make 55 mL of solution.
       (b)   Dissolve 0.35 g of K_2HPO_4 in enough water to make 10 mL of solution.
       (c)   Dissolve 0.33 g of $(NH_4)_2SO_4$ in enough water to make 0.25 liters of solution.

8.   (a) 1) 4.0×10^{-6} M $\dfrac{11\ \mu g}{10\ mL} \times \dfrac{1\ mol}{272 \times 10^6\ \mu g} = \dfrac{1000\ mL}{1\ L}$

       (b) 4.8×10^{-4} M   (c) 6.0×10^{-3} M   (d) 1.2×10^{-5} M

10. 2500 mL $\quad\quad\quad 2.5 \cancel{g} \times \dfrac{100 \text{ mL}}{0.1 \cancel{g}} = 2500 \text{ mL}$

12. (a) Dissolve 25 g of KOH in enough water to make 0.50 liter of solution.
    (b) Dissolve 10.3 g of glucose in enough water to make 125 mL of solution.

14. (a) 0.11 mg% $11 \, \mu g \times \dfrac{1 \text{ mg}}{1 \times 10^3 \, \mu g}$; $\dfrac{1.1 \times 10^{-2} \text{ mg}}{10 \text{ mL}} = \dfrac{? \text{ mg}}{100 \text{ mL}}$

    (b)   2.0 mg%   (c)   39 mg%   (d)   0.13 mg%

16. (a)   50 mL   (b)   1500 mL   (c)   50 mL
    (d)   10 mL   (e)   103 mL   (f)   530 mL

18. (a) 1.1 ppm $\dfrac{0.11 \text{ mg}}{100 \text{ mL}} = \dfrac{? \text{ mg}}{1000 \text{ mL}}$

    (b)   20 ppm   (c)   390 ppm   (d)   1.3 ppm

20. 0.24 mg Mg^{2+} $\quad 10 \cancel{\text{ mL}} \times \dfrac{2.0 \cancel{\text{ mEq}}}{1000 \cancel{\text{ mL}}} \times \dfrac{12 \text{ mg}}{1 \cancel{\text{ mEq } Mg^{2+}}}$

22. Yes, the blood potassium level is 2.0 mEq/liter

    $\dfrac{3.9 \cancel{\text{ mg } K^+}}{50 \cancel{\text{ mL}}} \times \dfrac{1 \text{ mEq}}{39.1 \cancel{\text{ mg } K^+}} \times \dfrac{1000 \cancel{\text{ mL}}}{1 \text{ L}}$

24 (a) 0.6% KCl   (b) 6.4 mg% urea

26. (a) 45 mL of stock solution plus enough water to make 55 mL of solution.
    (b) 260 mL of stock solution plus enough water to make 650 mL of solution.

28. The freezing point of the ocean is lower than that of fresh water and the boiling point is higher because of the dissolved salts.

30. 0.3 osm/liter $\dfrac{0.9 \cancel{g} \text{ NaCl}}{100 \cancel{\text{ mL}}} \times \dfrac{1 \text{ mol}}{58.5 \cancel{g}} \times \dfrac{1000 \cancel{\text{ mL}}}{1 \text{ L}} = 0.15 \text{ M}$

32. Water is hypotonic to the prune cells. The water molecules will enter the prune cells and cause them to swell. The cell membrane is also permeable to sugar molecule and they will migrate from the cell to a region of lower concentration, the water.

281

34. 1.2 osm/L

36. (c) 0.1 M NaOH

38. (a) hypertonic   (b) isotonic   (c) hypotonic

40.
(a) 2.09 μg

$$12 \, \cancel{oz} \times \frac{1 \, \cancel{qt}}{32 \, \cancel{oz}} \times \frac{1 \, L}{1.06 \, \cancel{qt}}$$

$$\frac{5.90 \, \mu g}{1 \, \cancel{L}} \times 0.354 \, \cancel{L}$$

(b) 2.02 μg

$$\frac{0.200 \, \mu g}{1 \, \cancel{L}} \times 0.354 \, \cancel{L}$$

42.
18 g Na^+

$$5.5 \, \cancel{L} \times \frac{142 \, \cancel{mEq}}{1 \, \cancel{L}} \times \frac{23 \, \cancel{mg}}{1 \, \cancel{mEq}} \times \frac{1 \, g}{1000 \, \cancel{mg}}$$

44. The increase in protein concentration in the cells would cause the fluid levels to increase and the tissues to swell.

46.

	glucose	NaCl	$NaHCO_3$	KCl
(a)	2.00%	0.35%	0.25%	0.15%
(b)	0.111 M	0.060 M	0.030 M	0.020 M

(c) 0.331 osmol/liter (add the osmolarities of the solutes)

Chapter 11

2. (a) $\underset{\text{acid}_1}{HSO_4^-} + \underset{\text{base}_2}{H_2O} \rightleftharpoons \underset{\text{acid}_2}{H_3O^+} + \underset{\text{base}_1}{SO_4^{2-}}$

(b) $\underset{\text{base}_1}{NH_2^-} + \underset{\text{acid}_2}{H_2O} \rightleftharpoons \underset{\text{acid}_1}{NH_3} + \underset{\text{base}_2}{OH^-}$

4. Bases form solutions in water that taste bitter, feel slippery to the touch, and turn litmus paper from red to blue.

6. (a) HPO_4^{2-} (b) HSO_3^- (c) ClO^-
   (d) NH_3 (e) OH^- (f) NH_2^-

12. $H_3ASO_4 + 3NaOH \rightleftharpoons 3H_2O + Na_3AsO_4$

$$3H^+ + 3OH^- \rightleftharpoons 3H_2O$$

14.   (a) 1×10^{-8} (c) 4.0×10^{-12}
        (b) 1.8×10^{-12} (d) 2.8×10^{-10}

16.   12, 9.5, 8.4, 7.4   7   6.3, 5.5, 4, 1.4
        basic   neutral   acidic

18.   (a) acidic   (b) basic   (c) basic

20.   (a) solution A is more acidic

        (b) solution A $[H^+] = 1 \times 10^{-3}$ M   solution B $[H^+] = 1 \times 10^{-5}$ M

        (c) The two hydrogen ion concentrations differ by a factor of 100 (10^2).

22.   carbonic acid/bicarbonate

24.   The following equilibrium exists in the buffer system:

$$H_2CO_3 \rightleftharpoons HCO_3^- + H^+$$

        (a)   When a acid is added to the system, the equilibrium will shift to the left, thus lowering the H^+ concentration.
        (b)   When a base is added to the system, the equilibrium will shift to the right, to replace the H^+ that reacts with the base.

26.   The baking soda dissolves to form the sodium ion and the bicarbonate ion. In the stomach the bicarbonate ion acts as a base to neutralize stomach acid.
$$HCO_3^- + H^+ \rightleftharpoons H_2CO_3 \rightleftharpoons CO_{2(g)} + H_2O$$

28.   Calcium hydroxide is a base and would help to neutralize the acidity of the lakes.

30.   Carbon dioxide dissolves in water to form the weak acid carbonic acid. Carbonic acid will dissociate to form hydrogen ions which lower the pH from the neutral 7 of pure water.

Integrated Problems

2.   Aluminum is a good conductor so the outside of the cup becomes cold very quickly.

When water molecules in the air collide with the cold surface of the cup, they condense. The outside of the cups made of glass or ceramic do not become as cold as quickly and therefore, do not have the condensation occur.

4. 1500 mL of 10.0% dextran

$$165 \text{ lb} \times \frac{1 \text{ kg}}{2.2 \text{ lb}} \times \frac{2.00 \text{ g dextran}}{1 \text{ kg}} \times \frac{100 \text{ mL}}{10.0 \text{ g dextran}}$$

6. Ammonia will diffuse from the blood into the kidney tubules. The hydrogen ions present in the acidic urine will shift the equilibrium between NH_3 and NH_4^+ toward the formation of NH_4^+. NH_4^+ cannot diffuse back into the blood and will be excreted in the urine.

8. Yes, mercury

As: $\dfrac{40 \text{ μg}}{\text{liter}} \times \dfrac{1 \text{ mg}}{1000 \text{ μg}} = \dfrac{0.04 \text{ mg}}{\text{liter}}$

Cd, Hg: $\dfrac{5 \text{ μg}}{\text{liter}} \times \dfrac{1 \text{ mg}}{1000 \text{ μg}} = \dfrac{0.005 \text{ mg}}{\text{liter}}$

10. $H^+ + F^- \rightleftharpoons HF$

The stomach is very acidic (pH < 3) and these conditions favor the formation of hydrofluoric acid. The intestines are basic (pH > 7) and these conditions favor the formation of fluoride ion. Therefore, hydrofluoric acid will be absorbed in the stomach and the fluoride ion in the small intestines.

12. (a) 0.43 osmol/liter, it is hypertonic

5% dextrose: $\dfrac{5 \text{ g}}{100 \text{ mL}} \times \dfrac{1000 \text{ mL}}{1 \text{ L}} \times \dfrac{1 \text{ mol}}{180 \text{ g}} = \dfrac{0.28 \text{ mol}}{1 \text{ L}}$

0.45% NaCl : $\dfrac{0.45 \text{ g}}{100 \text{ mL}} \times \dfrac{1000 \text{ mL}}{1 \text{ L}} \times \dfrac{1 \text{ mol}}{58.5 \text{ g}} = \dfrac{0.077 \text{ mol}}{1 \text{ L}}$

5% dextrose in 0.45% NaCl = 0.28 osmol/L + 2(0.077 osmol/L)
= 0.43 osmol/L

0.9% NaCl = 0.154 M = 0.31 osmol/L

(b) Yes

Chapter 12

2. Molecules found in living organisms are mainly long chains, branched chains, or rings which have carbon atoms as the axis or backbone of the molecule. Carbon is able to form strong stable bonds with up to four other carbon atoms, and also with atoms of elements such as oxygen, hydrogen, sulfur, phosphorus, and nitrogen.

4. A functional group is a particular arrangement of atoms that gives rise to a particular set of reactions.

6. sp, sp^2, sp^3 carbon uses sp^3 hybrid orbitals in methane

8. $CH_3CH_2CH_2CH_2CH_2CH_3$

$CH_3CH_2CHCH_2CH_3$
$\quad\quad\quad |$
$\quad\quad\quad CH_3$

$\quad\quad CH_3$
$\quad\quad |$
$CH_3\text{-}C\text{-}CH_2CH_3$
$\quad\quad |$
$\quad\quad CH_3$

$CH_3CHCH_2CH_2CH_3$
$\quad |$
$\quad CH_3$

$\quad\quad CH_3\;\; CH_3$
$\quad\quad |\quad\;\; |$
$CH_3 - CH - CH - CH_3$

10. (a) $(CH_3)_2CHCH_2CH_3$ (b) $(CH_3)_2CHCH(CH_3)CH(CH_3)CH_2CH_3$
    (c) $(CH_3)_4C$

12. In organic chemistry the term saturated means that each carbon has four single bonds, no additional atoms can be added to the molecule. In solutions, the term saturated means that the solution hold all of the solute that it can under the current conditions.

14. (a)   same   (b)   same   (c)   different

16. (a) $CH_3(CH_2)_3\text{-}CH\text{-}(CH_2)_4CH_3$ (d)   Cl   Cl
    $\quad\quad\quad\quad\quad |$ $\quad\quad\quad\quad |\quad |$
    $\quad\quad\quad\quad\quad CH_2CH_3$ $\quad\quad Cl\text{-}CH\text{-}CH\text{-}CH_2CH_3$

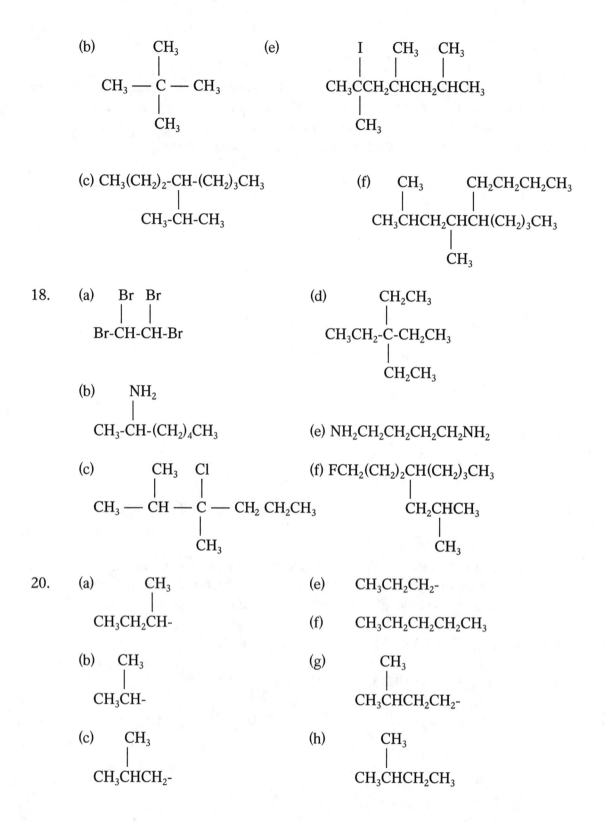

(b)

CH₃
|
CH₃ — C — CH₃
|
CH₃

(e)

$$\underset{\underset{CH_3}{|}}{\overset{\overset{I}{|}}{CH_3C}}CH_2\underset{}{\overset{\overset{CH_3}{|}}{CH}}CH_2\overset{\overset{CH_3}{|}}{CH}CH_3$$

(c) CH₃(CH₂)₂-CH-(CH₂)₃CH₃
|
CH₃-CH-CH₃

(f)

$$CH_3\overset{\overset{CH_3}{|}}{CH}CH_2\underset{\underset{CH_3}{|}}{\overset{\overset{CH_2CH_2CH_2CH_3}{|}}{CH}CH}(CH_2)_3CH_3$$

18. (a)   Br Br
| |
Br-CH-CH-Br

(d)

$$CH_3CH_2-\underset{\underset{CH_2CH_3}{|}}{\overset{\overset{CH_2CH_3}{|}}{C}}-CH_2CH_3$$

(b)

NH₂
|
CH₃-CH-(CH₂)₄CH₃

(e) NH₂CH₂CH₂CH₂CH₂NH₂

(c)

$$CH_3 — \overset{\overset{CH_3}{|}}{CH} — \underset{\underset{CH_3}{|}}{\overset{\overset{Cl}{|}}{C}} — CH_2\ CH_2CH_3$$

(f) FCH₂(CH₂)₂CH(CH₂)₃CH₃
|
CH₂CHCH₃
|
CH₃

20. (a)

CH₃
|
CH₃CH₂CH-

(e)   CH₃CH₂CH₂-

(f)   CH₃CH₂CH₂CH₂CH₃

(b)   CH₃
|
CH₃CH-

(g)

CH₃
|
CH₃CHCH₂CH₂-

(c)   CH₃
|
CH₃CHCH₂-

(h)

CH₃
|
CH₃CHCH₂CH₃

286

(d)

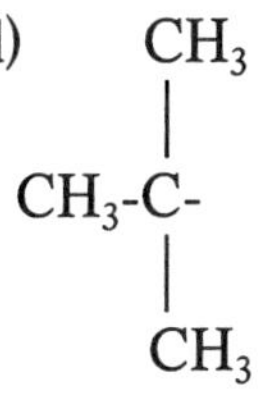

$$CH_3-\underset{\underset{\textstyle CH_3}{|}}{\overset{\overset{\textstyle CH_3}{|}}{C}}-$$

22. 1-bromo-2,2dimethylbutane: the longest carbon chain has 4 carbons

24. (a) $CH_4 + Br_2 \longrightarrow CH_3Br + HBr$

(b) $C_3H_8 + Br_2 \longrightarrow C_3H_7Br + HBr$ ($CH_3CH_2CH_2Br$, $CH_3CHBrCH_3$)

26. The alkenes and alkynes are described as being unsaturated because additional atoms can be added to these molecules.

28. (a) 3-methyl-2-pentene (c) 3-methyl-1,4-pentadiene
    (b) 3,3-dimethyl-1-hexene (d) 2,3,4-trichloro-2-hexene

30. (a) sp^2-sp^2-sp^2-sp-sp^2-sp^3 (b) sp^2-sp^2-sp^2-sp^2-sp^3

32.

(a)

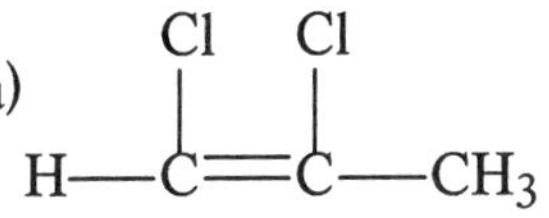

$$H-\underset{\overset{|}{Cl}}{C}=\underset{\overset{|}{Cl}}{C}-CH_3$$

(b)

$$CH_3-\underset{\overset{|}{H}}{C}=\underset{\overset{|}{CH_3}}{\overset{\overset{|}{H}}{C}}-CH-\bigcirc$$

(c)

$$CH_3CH_2-\overset{\overset{\textstyle H}{|}}{C}=\overset{\overset{\textstyle H}{|}}{C}-CH_2CH_3$$

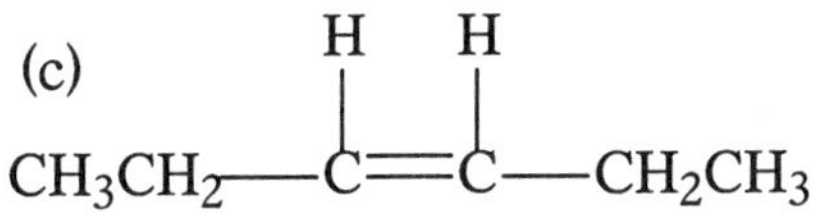

(d)

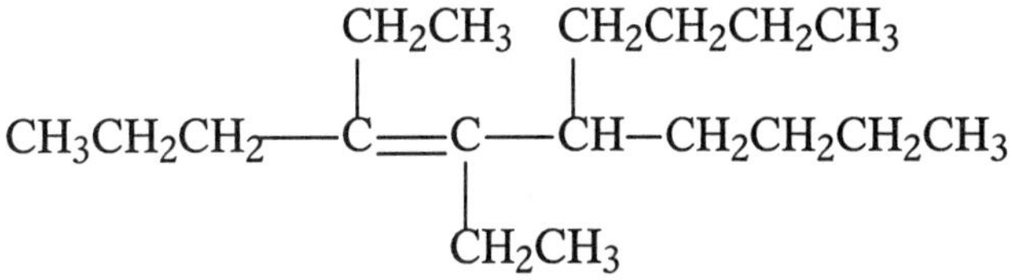

$$CH_3CH_2CH_2-\underset{\overset{|}{CH_2CH_3}}{C}=\overset{\overset{\textstyle CH_2CH_3}{|}}{C}-\overset{\overset{\textstyle CH_2CH_2CH_2CH_3}{|}}{CH}-CH_2CH_2CH_2CH_3$$

34.

(a)

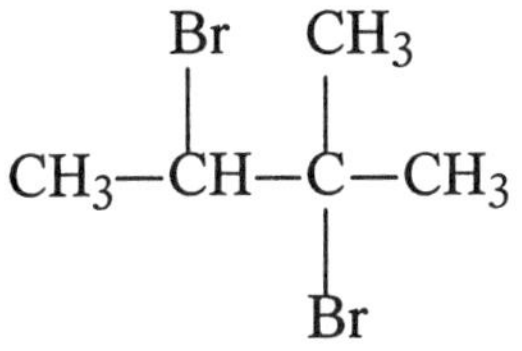

$$CH_3-\underset{\overset{|}{Br}}{\overset{\overset{|}{Br}}{CH}}-\underset{\overset{|}{Br}}{\overset{\overset{|}{CH_3}}{C}}-CH_3$$

(c)

$$H_3C-CH\underset{\overset{|}{OH}}{\cdot}\underset{\overset{|}{OH}}{\overset{\overset{|}{CH_3}}{C}}-CH_3$$

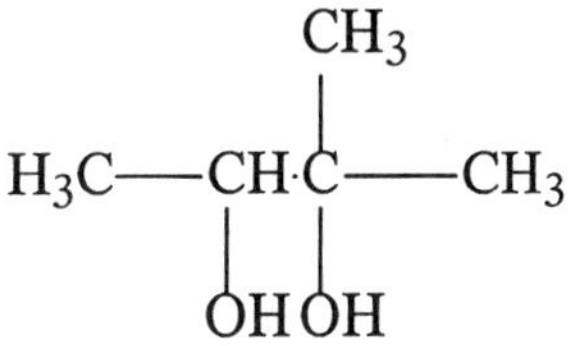

(b)

$$CH_3-CH_2-CH_2-\underset{\overset{|}{OH}}{CH}-CH_2-CH_3$$

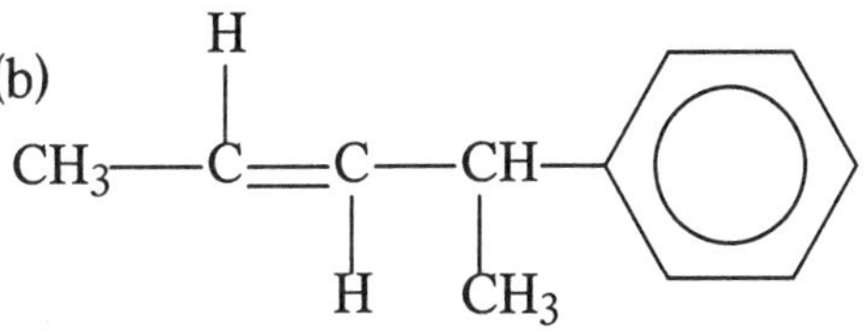

287

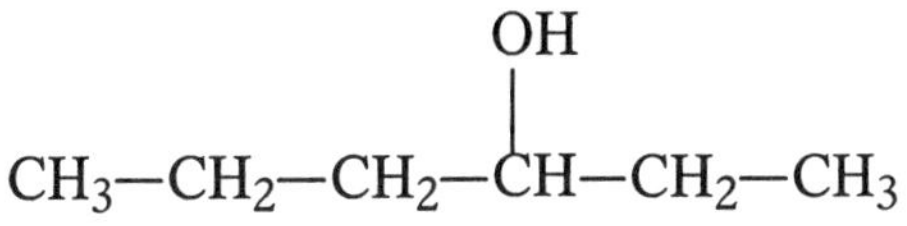

36.

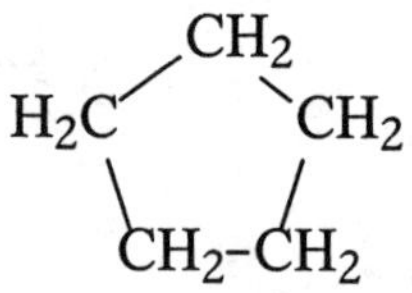

$$\begin{array}{c} \text{CH}_2 \\ \text{H}_2\text{C} \qquad \text{CH}_2 \\ \text{CH}_2\text{--CH}_2 \end{array}$$

cyclopentane

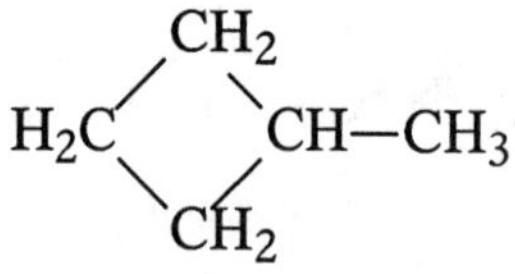

$$\begin{array}{c} \text{CH}_2 \\ \text{H}_2\text{C} \qquad \text{CH--CH}_3 \\ \text{CH}_2 \end{array}$$

methylcyclobutane

$$\begin{array}{c} \text{CH}_3 \\ \text{H}_2\text{C} \\ \qquad \text{C--CH}_3 \\ \text{H}_2\text{C} \end{array}$$

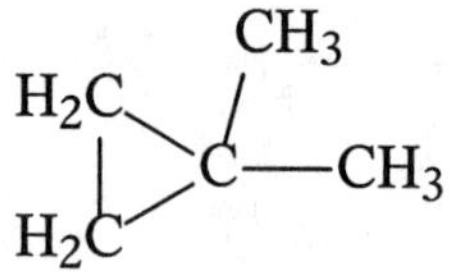

1,1-dimethylcyclopropan

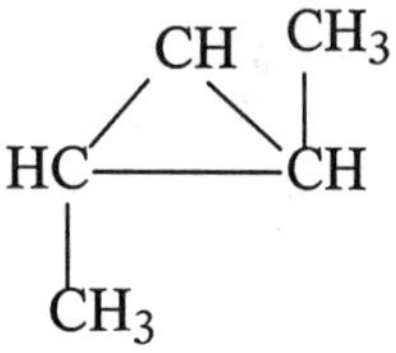

$$\begin{array}{c} \text{CH} \quad \text{CH}_3 \\ \text{HC} \text{----} \text{CH} \\ \text{CH}_3 \end{array}$$

trans-1,2-dimethylcyclopropane

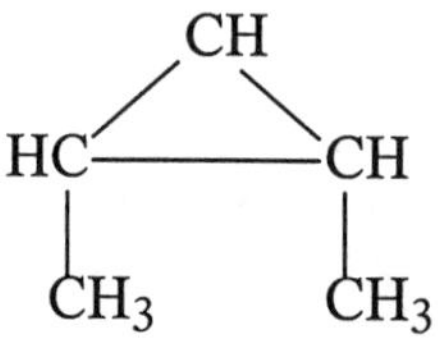

$$\begin{array}{c} \text{CH} \\ \text{HC} \text{----} \text{CH} \\ \text{CH}_3 \qquad \text{CH}_3 \end{array}$$

cis-1,2-dimethylcyclopropan

38.

(a) 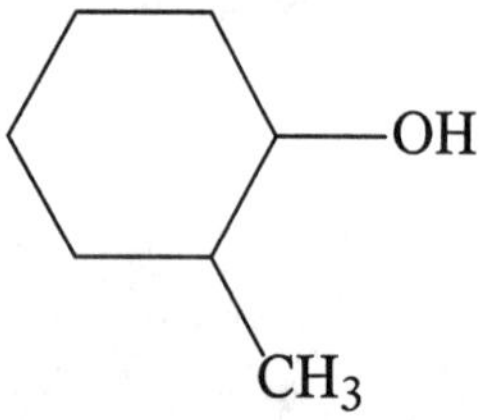 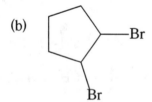 —OH or a structure —OH with CH₃

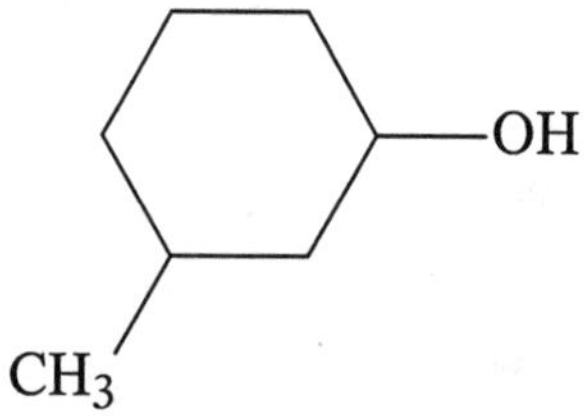

(b) a structure —Br with Br

(c) 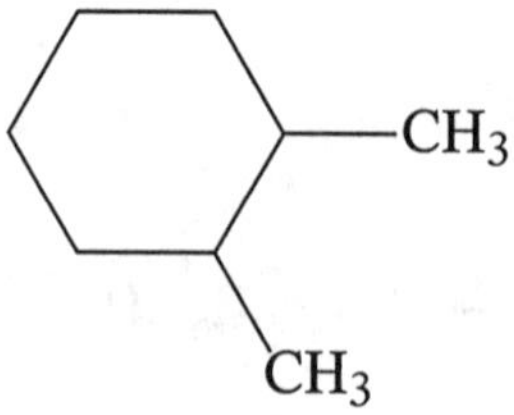 a structure —CH₃ with CH₃

40. Aromatic compounds have a resonance hybrid structure in the ring that gives the ring extra stability. These ring structures will only undergo substitution reactions and not addition reactions.

42
(a) alkyne, unsaturated
(b) alkane, saturated
(c) alkane, saturated
(d) alkene, unsaturated
(e) alkene, unsaturated

44. $CH_3CH_2CHCl_2$ 1,1-dichloropropane

        $CH_3CHClCH_2Cl$ 1,2-dichloropropane

        $CH_3CCl_2CH_3$ 2,2-dichloropropane

        $CH_2ClCH_2CH_2Cl$ 1,3-dichloropropane

46.   (only the carbon chain is shown)

        C=C-C=C-C   conjugated   C=C-C=C   conjugated

        C-C=C=C-C |

        C=C-C-C=C   C

        C=C=C-C-C   C=C=C-C

                                                      |

                                                      C

48.   An electric spark might ignite the inflammable anesthetic that could have escaped into the room.

50.

    1-phenylpropane   2-phenylpropane   2-ethyltoluene

    3-ethyltoluene   4-ethyltoluene   2,3-dimethyltol

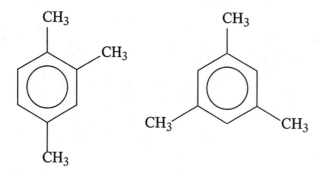

2,4-dimethyltoluene   3,5-dimethyltoluene

Chapter 13

2.   Compounds that participate in the chemical reactions of living organisms are said to be physiologically active.

4.   (a)   ethyl methyl ether   (c)   methyl pentyl ether
        (b)   cyclopentyl ethyl ether   (d)   diethyl ether

6.

(a) $CH_3-CH_2-\underset{\underset{CH_3}{|}}{\overset{\overset{OH}{|}}{C}}-CH_3$ (b) $CH_3-\underset{}{\overset{\overset{OH}{|}}{C}H}-CH_3$

(c) $CH_3-\underset{}{\overset{\overset{CH_3}{|}}{C}H}-CH_2-OH$

8.   (a)   2-methyl-2-butanol   (c)   3,3-dimethyl-1-pentanol
        (b)   cyclopentanol   (d)   2,3-dibromo-2-hexanol

10.   Ethanol has a higher boiling point than propane and methyl ether because of the hydrogen bonding that can form between the molecules of ethanol.

12.   (only the carbon chain and functional groups are shown)

        -C-O-C-C-C-, methyl propyl ether   -C-C-O-C-C-, diethyl ether

$$\underset{\begin{array}{c}\text{C}\\|\end{array}}{\text{-C-O-C-C-}}, \text{ isopropyl methyl ether} \qquad \text{-C-C-C-C-OH, 1-butanol}$$

$$\underset{\begin{array}{c}\text{OH}\\|\end{array}}{\text{-C-C-C-C-}}, \text{ 2-butanol} \qquad \underset{\begin{array}{c}\text{C}\\|\end{array}}{\text{-C-C-C-OH}}, \text{ 2-methyl-1-propanol}$$
(isobutyl alcohol)

$$\underset{\begin{array}{c}\text{C}\\|\end{array}}{\text{-C-C-C-OH}}, \text{ 1-methyl-1-propanol, (sec-butyl alcohol)}$$

14.

(a) $CH_3-CH_2-\underset{\begin{array}{c}\text{OH}\\|\end{array}}{CH}-CH_3$

(c) cyclohexane with —OH

(b) $CH_3-CH_2-CH_2-\underset{\begin{array}{c}\text{OH}\\|\end{array}}{CH}-CH_2-CH_3$

16.   (a) $CH_3CH_2CH=CH_2$ (b) $CH_3CH_2CH=CHCH_2CH_3$

18.

(a) cyclohexane with CH3 and —OH

(c) $CH_3-CH_2-CH_2-\overset{\begin{array}{c}\text{CH}_3\\|\end{array}}{\underset{\begin{array}{c}|\\\text{OH}\end{array}}{C}}-CH_3$

(b) $CH_3-CH_2-\overset{\begin{array}{c}\text{CH}_3\\|\end{array}}{CH}-\overset{\begin{array}{c}\text{CH}_3\\|\end{array}}{CH}-\overset{\begin{array}{c}\text{CH}_3\\|\end{array}}{CH}-CH_2-OH$

20.   (only the carbon chain and functional groups are shown)

$$\underset{\text{-C-C-C-C-C-H, pentanal}}{\overset{\displaystyle\overset{\text{O}}{\|}}{}}$$

$$\underset{\text{-C-C-C-C-H, 2-methylbutanal}}{\overset{\displaystyle\overset{\text{C O}}{|\ \|}}{}}$$

$$\underset{\text{-C-C-C-C-H, 3-methylbutanal}}{\overset{\displaystyle\overset{\text{C O}}{|\ \ \|}}{}}$$

$$\underset{\underset{\text{C}}{|}}{\underset{\text{-C-C-C-H, 2,2-dimethylpropanal}}{\overset{\displaystyle\overset{\text{C O}}{|\ \|}}{}}}$$

$$\underset{\text{-C-C-C-C-C-, 2-pentanone}}{\overset{\displaystyle\overset{\text{O}}{\|}}{}}$$

$$\underset{\text{-C-C-C-C-C-, 3-pentanone}}{\overset{\displaystyle\overset{\text{O}}{\|}}{}}$$

$$\underset{\text{-C-C-C-C-, 3-methyl-2-butanone}}{\overset{\displaystyle\overset{\text{C O}}{|\ \|}}{}}$$

22.   Diabetes mellitus

24.

(a) $CH_3-\overset{\displaystyle\overset{O}{\|}}{C}-OH$ (c) $CH_3-CH_2-CH_2-CH_2-CH_2-\overset{\displaystyle\overset{O}{\|}}{C}-OH$

(b) $CH_3-CH_2-CH_2-\overset{\displaystyle\overset{O}{\|}}{C}-OH$ (d)

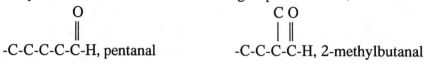

26.   (a) $\underset{\underset{OCH_3}{|}}{\overset{\overset{OH}{|}}{CH_3\text{-}C\text{-}H}}$ (b) $\underset{\underset{OCH_2CH_3}{|}}{\overset{\overset{OH}{|}}{CH_3\text{-}C\text{-}CH_3}}$

(c)

$$CH_3CH_2-\overset{\overset{\displaystyle OH}{|}}{\underset{\underset{\displaystyle OCH_3}{|}}{C}}-CH_3$$

(d)

$$H-\overset{\overset{\displaystyle OH}{|}}{\underset{\underset{\displaystyle OCH_2CH_3}{|}}{C}}-H$$

(e)

$$CH_3CH_2-\overset{\overset{\displaystyle OH}{|}}{\underset{\underset{\displaystyle OCH_3}{|}}{C}}-H$$

(f)

$$CH_3-\overset{\overset{\displaystyle OH}{|}}{\underset{\underset{\displaystyle OCH_2CH_2CH}{|}}{C}}-CH_3$$

28.

(a) $CH_3-(CH_2)_6-\overset{\overset{\displaystyle O}{\|}}{C}-OH$

(b) $CH_3-\overset{\overset{\displaystyle CH_3}{|}}{\underset{\underset{\displaystyle Cl}{|}}{C}}-\overset{\overset{\displaystyle CH_3}{|}}{CH}-\overset{\overset{\displaystyle O}{\|}}{C}-OH$

(c) $CH_3-CH_2-\overset{|}{CH}-CH_2-\overset{\overset{\displaystyle O}{\|}}{C}-OH$ (with phenyl group on CH)

(d) $HO-\overset{\overset{\displaystyle O}{\|}}{C}-CH_2-CH_2-\overset{\overset{\displaystyle O}{\|}}{C}-OH$

30. acetic acid(vinegar), oxalic acid (in rhubard), benzoic acid(a saltis used to retard spoilage), lactic acid (sour milk)

32.

(a) $CH_3-CH_2-CH_2-\overset{\overset{\displaystyle O}{\|}}{C}-OH$

(b) $CH_3-CH_2-CH_2-CH_2-\overset{\overset{\displaystyle O}{\|}}{C}-OH$

(c) $CH_3-CH_2-CH_2-\overset{\overset{\displaystyle CH_3}{|}}{CH}-\overset{\overset{\displaystyle CH_3}{|}}{CH}-\overset{\overset{\displaystyle O}{\|}}{C}-OH$

(d) $CH_3-CH_2-\overset{\overset{\displaystyle CH_3}{|}}{CH}-\overset{\overset{\displaystyle CH_3}{|}}{CH}-CH_2-\overset{\overset{\displaystyle O}{\|}}{C}-OH$

34. (a)

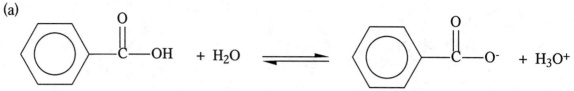

(b)

$$HCl + H_2O \rightleftharpoons H_3O^+ + Cl^-$$

(c) CH₃—CH—C—OH + H₂O $\rightleftharpoons$ CH₃—CH—C—O⁻ + H₃O⁺

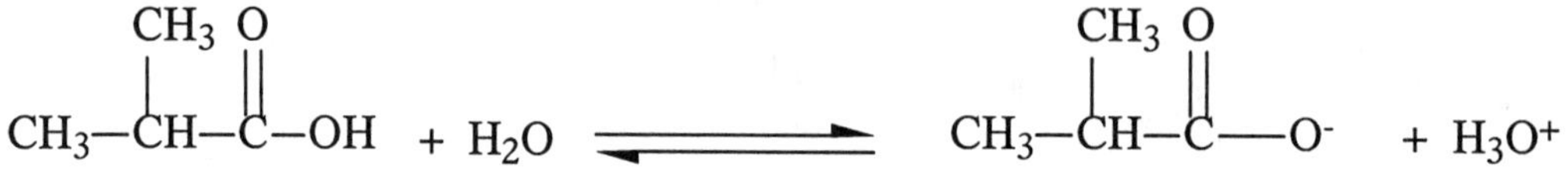

36.   The part of each ester that came from the carboxylic acid is bracketed.

(a)

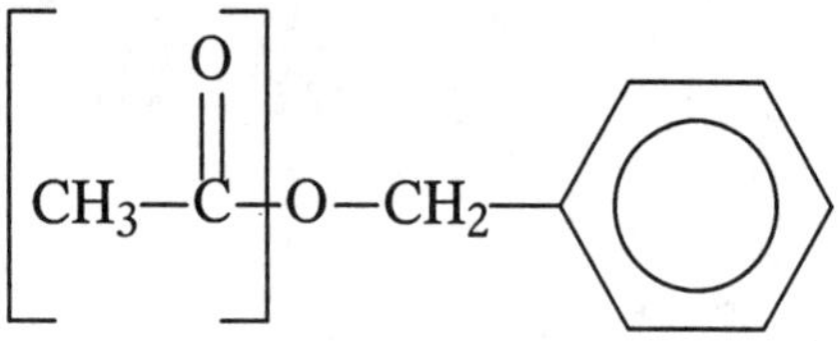

$$\left[CH_3-\overset{O}{\overset{\|}{C}}\right]O-CH_2-\bigcirc$$

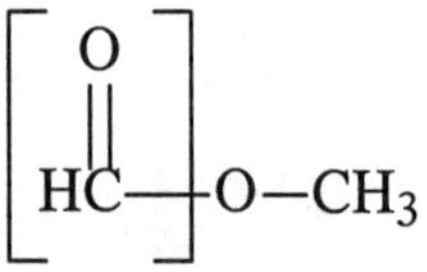

(b)

$$\left[H\overset{O}{\overset{\|}{C}}\right]O-CH_3$$

(c)

$$\left[CH_3-CH_2-CH_2-\overset{CH_3}{\underset{|}{CH}}-\overset{O}{\overset{\|}{C}}\right]O-\bigcirc$$

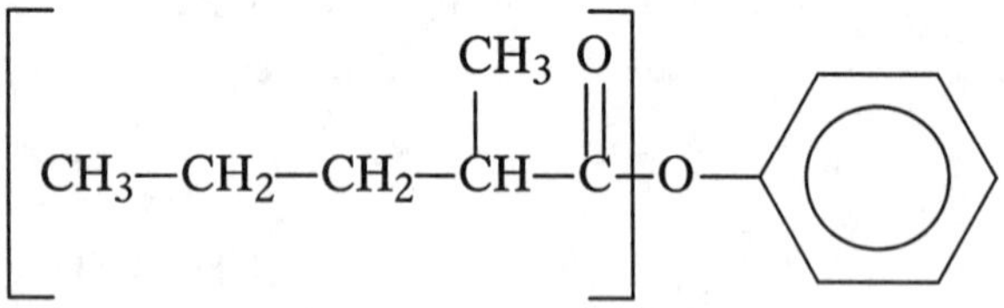

(d)

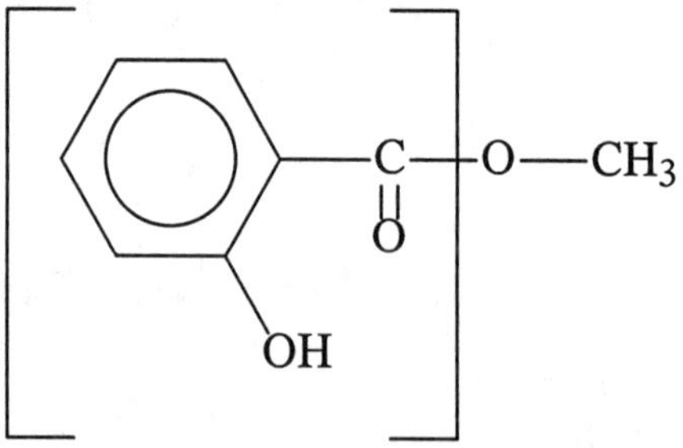

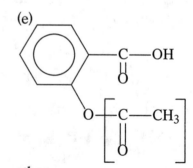

(e)

38.   (a)   carboxylic acid   (g)   ether
      (b)   ether   (h)   alcohol
      (c)   ester   (i)   aldehyde
      (d)   aldehyde   (j)   ketone
      (e)   ketone   (k)   ester
      (f)   alcohol, carboxylic acid   (l)   alcohol

294

40. acetylsalicylic acid (aspirin) is an analgesic and antipyretic, nitroglycerin relieves the pain of angina pectoris, methyl salicylate is used in ointments.

42. 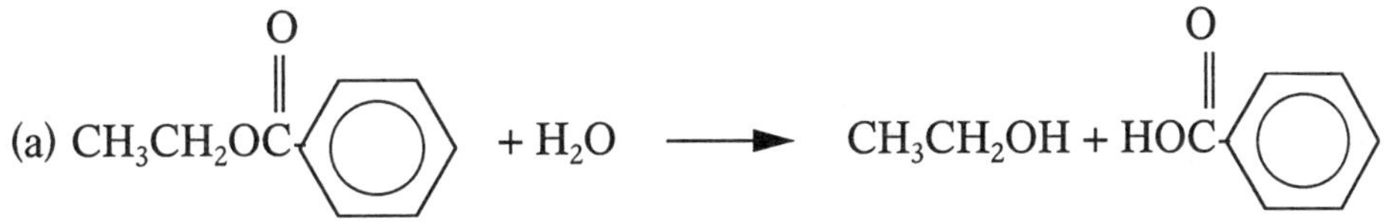 (a) $CH_3CH_2OC\overset{O}{\underset{\|}{}}$〈 〉 + H_2O ⟶ CH_3CH_2OH + $HOC\overset{O}{\underset{\|}{}}$〈 〉

　　　　　　　　　　　　　　　　　　　ethanol　　　　　benzoic acid

(b) $CH_3\underset{\underset{CH_3}{|}}{CH}OCCH_2CH_2CH_3$ + H_2O ⟶ $CH_3\underset{\underset{CH_3}{|}}{CH}OH$ + $HOCCH_2CH_2CH_3$

　　　　　　　　　　　　　　　　　　　　isopropyl　　　butanoic acid
　　　　　　　　　　　　　　　　　　　　alcohol

(c) $CH_3(CH_2)_6CH_2OCCH_3$ + H_2O ⟶ $CH_3(CH_2)_6CH_2OH$ + $HOCCH_3$

　　　　　　　　　　　　　　　　　　　octanol　　　　acetic acid

(d) $CH_3 - \underset{\underset{CH_3}{|}}{\overset{\overset{CH_3}{|}}{C}} - OCCH_2CH_3$ + H_2O ⟶ $CH_3-\underset{\underset{CH_3}{|}}{\overset{\overset{CH_3}{|}}{C}}-OH$ + $HOCCH_2CH_3$

　　　　　　　　　　　　　　　　　t-butyl　　　　propanoic acid
　　　　　　　　　　　　　　　　　alcohol

44. The O-H bond is polar with the oxygen at the negative end of the bond. The C=O bond is polar with the carbon at the positive end of the bond. The partially negative oxygen is attracted to the partially negative carbon.

46. The student can use the Benedict's test. The aldehyde(pentanal) will giva positive test. A brick red precipitate will form. 2-pentanol will not react.

48. Estrone is converted to estradiol by the reduction of the ketone group.

Chapter 14

2.

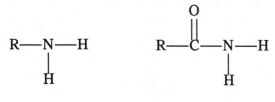

R—N—H
    |
    H

amine

$$R—\overset{\displaystyle O}{\overset{\|}{C}}—N—H$$
             |
             H

amide

4. (a) tert-butylamine   (d) isopropylmethylamine
   (b) N-ethyl-N-methylaniline   (e) dimethylethylamine
   (c) propylamine   (f) isopentylamine

6.

(a) 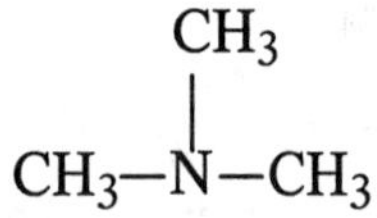
$$CH_3—\overset{\displaystyle CH_3}{\overset{|}{N}}—CH_3$$

$$CH_3—\overset{\displaystyle O}{\overset{\|}{C}}—OH$$

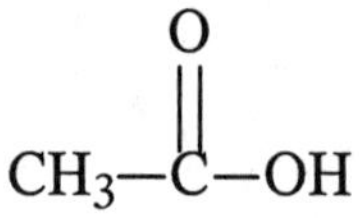

(b) $CH_3\!\!-\!\!-\!\!CH_2—NH_2$

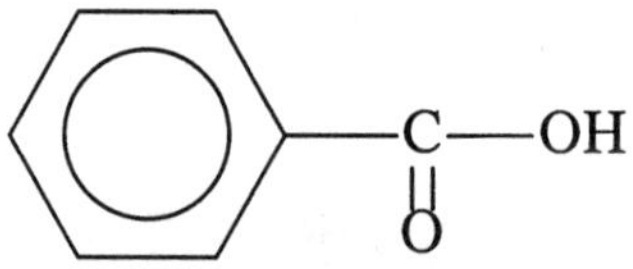

benzene ring—$\overset{\displaystyle }{\underset{\displaystyle O}{\overset{\displaystyle }{C}}}$—OH

(c) $CH_3—CH_2—NH—CH_3$

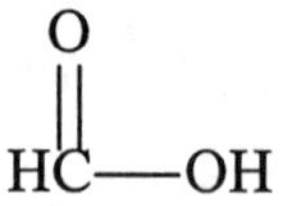

$$H\overset{\displaystyle O}{\overset{\|}{C}}\!\!-\!\!OH$$

8.

(a) 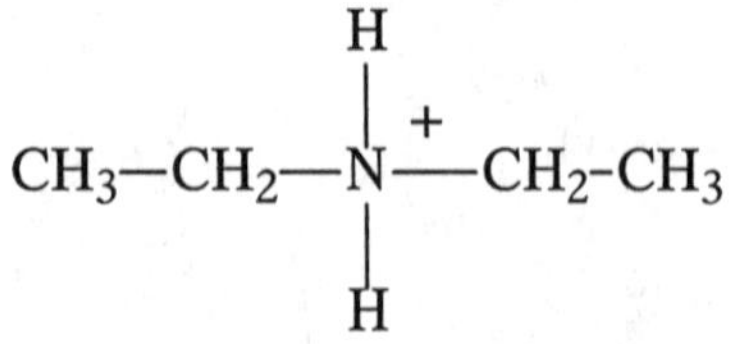
$$CH_3—CH_2—\overset{\displaystyle H}{\underset{\displaystyle H}{\overset{|}{\overset{+}{N}}}}—CH_2\text{-}CH_3 \quad Br^-$$

(b) 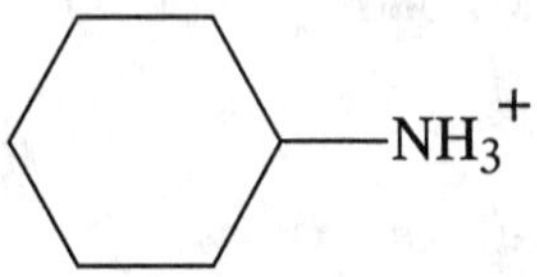 cyclohexane—NH_3^+ Cl^-

(c) 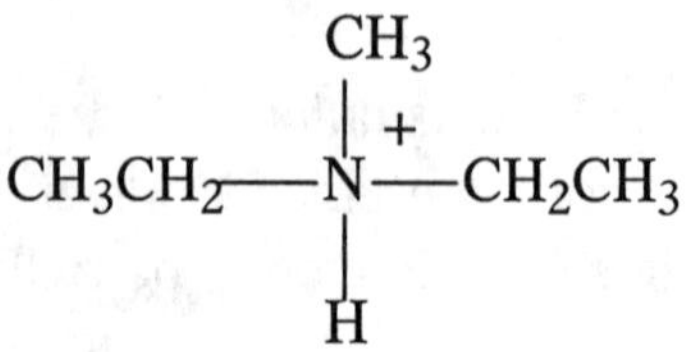
$$CH_3CH_2—\overset{\displaystyle CH_3}{\underset{\displaystyle H}{\overset{|}{\overset{+}{N}}}}—CH_2CH_3 \quad Br^-$$

10.

(a)

$$\text{C}_6\text{H}_5\text{-CH}_2\overset{\overset{\displaystyle \text{CH}_3}{|}}{\text{NH}}$$

(d) $\text{CH}_3\text{CH}_2\text{CH}_2\text{CH}_2\text{CH}_2\overset{\overset{\displaystyle \text{O}}{\|}}{\text{C}}\text{-N}\overset{\overset{\displaystyle \text{CH}_2\text{CH}_3}{|}}{}\text{CH}_2\text{CH}_3$

(b) $\text{CH}_3\text{CH}_2\text{CH}_2\text{CH}_2\overset{\overset{\displaystyle \text{O}}{\|}}{\text{C}}\text{NH}_2$

(e) $\text{CH}_3\overset{\overset{\displaystyle \text{NHCH}_3}{|}}{\text{CH}}\text{CH}_2\text{OH}$

(c) $\text{CH}_3\overset{\overset{\displaystyle \text{CH}_3}{|}}{\text{CH}}\text{- N -}\overset{\overset{\displaystyle \text{CH}_3}{|}}{\text{CHCH}_3}$, with CH_3CHCH_3 below N

(f) $\text{CH}_3\text{CH}_2\overset{\overset{\displaystyle \text{O}}{\|}}{\text{C}}\text{-N}\overset{}{\underset{\underset{\displaystyle \text{CH}_3}{|}}{}}\text{C}_6\text{H}_5$

12. Procaine is a tertiary amine and an ester. Lidocaine is a tertiary amine and an amide. they both contain an aromatic ring.

14. (a) pentanoic acid and ammonia
    (b) dimethylamine and acetic acid
    (c) aniline and propanoic acid
    (d) ethylmethylamine and hexanoic acid

16. They defend the plants against predators.

18. Primary amine

20.

$$\overset{\overset{\displaystyle \text{O}}{\|}}{\text{HCOH}} + \text{NH}_3 \longrightarrow \overset{\overset{\displaystyle \text{O}}{\|}}{\text{HCO}^-} + \text{NH}_4^+$$

22. This structure will repeat many times to form the polymer.

$$-\text{NH}-(\text{CH}_2)_6-\text{NH}-\overset{\overset{\displaystyle \text{O}}{\|}}{\text{C}}-(\text{CH}_2)_4-\overset{\overset{\displaystyle \text{O}}{\|}}{\text{C}}-$$

24. The drug is more soluble in the form of its salt.

Integrated problems

2. Acetic acid is a weak acid and along with its conjugate base, the acetate ion, forms a buffer that will protect a solution from changes in pH.

4. $\dfrac{1.2 \text{ mol Br}_2}{1000 \text{ mL}} \times 41.7 \text{ mL} = 0.05 \text{ mol Br}_2$

   $2.05 \text{ g} \times \dfrac{1 \text{ mol C}_6\text{H}_{10}}{82 \text{ g}} = 0.025 \text{ mol C}_6\text{H}_{10}$

   $\dfrac{0.05 \text{ mol Br}_2}{0.025 \text{ mol C}_6\text{H}_{10}} = 2 \text{ mol Br}_2 \text{ 2 double bonds}$

   $H_2C=CH-CH=CH-CH_2-CH_3$

6. Methanol is highly toxic, less than 30 mL can cause death. Methanol is oxidized to formaldehyde in the liver. Formaldehyde in the blood damages the tissues of the retina of the eye and can cause blindness. The formaldehyde is further oxidized to formic acid causing acidosis and a disruption of the body's ability to transport oxygen which can result in death.

8. (a) The nitrogen in an amine has a pair of unshared electrons that it can donate to a hydrogen ion thereby acting as a hydrogen acceptor or base.

   (b) $CH_3CH_2CH_2NH_2 + H_2O \longrightarrow CH_3CH_2CH_2NH_3^+ + OH^-$

   (c) dipropylamine. Basic strength increases with the length of the carbon chain.

10.
$$\underset{\displaystyle CCl_3CH}{\overset{\displaystyle O}{\overset{\|}{}}} + HOH \longrightarrow \underset{\displaystyle OH}{\overset{\displaystyle OH}{\underset{|}{\overset{|}{CCl_3CH}}}}$$

12. (a) $\overset{O}{\overset{\|}{HCOH}}$
    (b) methanoic acid
    (c) $4 \times 10^{11} \text{ mol}, 2 \times 10^7 \text{ ton}$

$$2 \times 10^{13} \, \cancel{g} \times \frac{1 \text{ mol}}{46 \, \cancel{g}} = 4 \times 10^{11} \text{ mol}$$

$$2 \times 10^{13} \, \cancel{g} \times \frac{1 \text{ lb}}{454 \, \cancel{g}} \times \frac{1 \text{ ton}}{2000 \, \cancel{lb}} = 2 \times 10^7 \text{ ton}$$

Chapter 15

2. (a) aldose, pentose   (c) ketose, hexose
    (b) ketose, heptose   (d) aldose, triose

4. The difference between starch (a digestible glucose polymer) and cellulose (an indigestible glucose polymer) is the position of the hydroxyl group on carbon 1 of the glucose molecule.

6. 32 aldoheptoses, 16 in the D-family and 16 in the L-family.

8.

10. milk and any milk product, like cheese, ice cream, and butter

12. β-Ribose has a hydroxyl group on carbon 2, but β-deoxyribose does not.

14.

299

16.

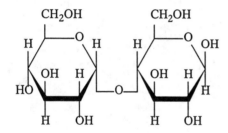

18.

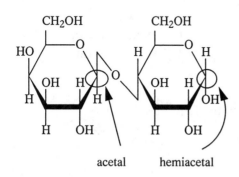

acetal   hemiacetal

20.

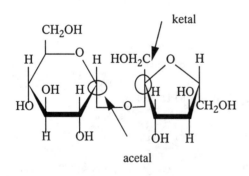

22.

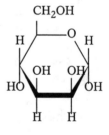

24.

26.   Starch is the storage form of glucose in plants.

28. Glycogen consists of straight chains of glucose joined by α,1:4 linkages with branching that occurs due to α,1:6 linkages. Its structureis very similar to that of amylopectin which is also branched. Amylose is not branched.

30. Humans cannot digest cellulose because we do not have the enzymes necessary to catalyze the hydrolysis of the β,1:4 linkages between the glucose units in cellulose.

32. glucose

34. (a) $Energy + 6\,CO_2 + 6\,H_2O \longrightarrow C_6H_{12}O_6 + 6\,O_2$
    (b) sunlight

36. Glucose is a monosaccharide normally found in the blood, but sucrose, a disaccharide, cannot be used by body cells and is normally hydrolyzed by enzymes in the digestive tract.

38. (a) glucose, glucose
    (b) Yes, there is a free aldehyde on the second glucose.
    (c) It is an α,1:6 linkage.
    (d) This linkage is found in molecules of glycogen and amylopectin.

40.

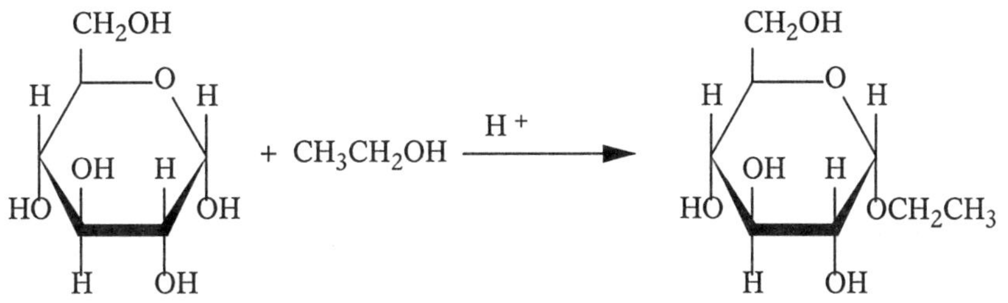

42. The glucosidic linkage would form between the alcohol groups on carbon 1 on both molecules of glucose.

44. (a)

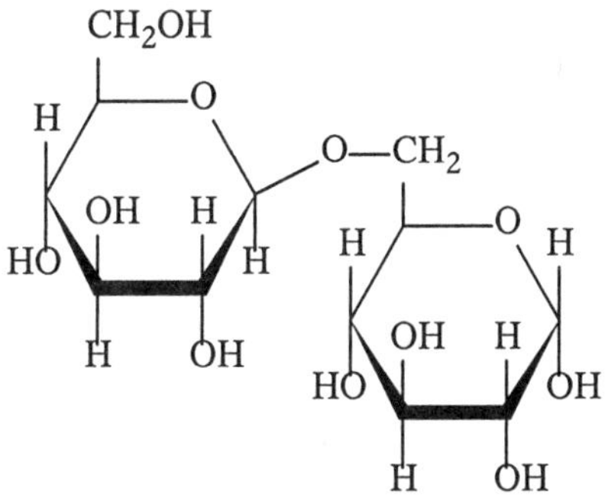

(b) Yes, it is a reducing sugar.

Chapter 16

2. A simple lipid yields glycerol and three fatty acids upon hydrolysis. A compound lipid yields an alcohol, fatty acids, and other compounds upon hydrolysis.

4.

(a)
$$CH_2OC(CH_2)_{14}CH_3$$ (with O double-bonded above the C)
$$CHOC(CH_2)_{14}CH_3$$ (with O double-bonded above the C)
$$CH_2OC(CH_2)_{14}$$ (with O double-bonded above the C)

(b)
$$CH_2OC(CH_2)_{18}CH_3$$ (with O double-bonded above the C)
$$CHOC(CH_2)_{16}CH_3$$ (with O double-bonded above the C)
$$CH_2OC(CH_2)_7CH=CH(CH_2)_7CH_3$$ (with O double-bonded above the C)

(c)
$$CH_2OC(CH_2)_7CH=CH(CH_2)_7CH_3$$ (with O double-bonded above the C)
$$CHOC(CH_2)_7CH=CH(CH_2)_7CH_3$$ (with O double-bonded above the C)
$$CH_2OC(CH_2)_7CH=CH(CH_2)_7CH_3$$ (with O double-bonded above the C)

6. The triacylglycerols in animal fats contain more saturated fatty acids than unsaturated fatty acids, whereas the triacylglycerols in vegetable oils contain more unsaturated fatty acids.

8

(a) $CH_3CH_2-\overset{H}{\underset{}{C}}=\overset{H}{\underset{}{C}}-CH_2-\overset{H}{\underset{}{C}}=\overset{H}{\underset{}{C}}-CH_2-\overset{H}{\underset{}{C}}=\overset{H}{\underset{}{C}}-(CH_2)_7COOH$

(b) $CH_3(CH_2)_4-\overset{H}{\underset{}{C}}=\overset{H}{\underset{}{C}}-CH_2-\overset{H}{\underset{}{C}}=\overset{H}{\underset{}{C}}-CH_2-\overset{H}{\underset{}{C}}=\overset{H}{\underset{}{C}}-CH_2-\overset{H}{\underset{}{C}}=\overset{H}{\underset{}{C}}-(CH_2)_3COOH$

10. (a) 10:0 (b) 15:1n-7

12. Linolenic acid is one of the essential fatty acids that cannot be produced by the body but that are essential for good health.

14. Waxes are insoluble because of their long hydrocarbon chain which is nonpolar.

16. (a) a　　　　　　　(b) a　　　　　(c) c

18.

(b)
$$\text{CH}_2\text{OC}(\text{CH}_2)_{18}\text{CH}_3$$ (with C=O)
|
$$\text{CHOC}(\text{CH}_2)_{16}\text{CH}_3$$ (with C=O) $+ \text{H}_2$
|
$$\text{CH}_2\text{OC}(\text{CH}_2)_7\text{CH}=\text{CH}(\text{CH}_2)_7\text{CH}_3$$ (with C=O)

$\longrightarrow$

$$\text{CH}_2\text{OC}(\text{CH}_2)_{18}\text{CH}_3$$
|
$$\text{CHOC}(\text{CH}_2)_{16}\text{CH}_3$$
|
$$\text{CH}_2\text{OC}(\text{CH}_2)_{16}\text{CH}_3$$

(c)
$$\text{CH}_2\text{OC}(\text{CH}_2)_7\text{CH}=\text{CH}(\text{CH}_2)_7\text{CH}_3$$ (with C=O)
|
$$\text{CHOC}(\text{CH}_2)_7\text{CH}=\text{CH}(\text{CH}_2)_7\text{CH}_3$$ (with C=O) $+ \text{H}_2$
|
$$\text{CH}_2\text{OC}(\text{CH}_2)_7\text{CH}=\text{CH}(\text{CH}_2)_7\text{CH}_3$$ (with C=O)

$\longrightarrow$

$$\text{CH}_2\text{OC}(\text{CH}_2)_{16}\text{CH}_3$$
|
$$\text{CHOC}(\text{CH}_2)_{16}\text{CH}_3$$
|
$$\text{CH}_2\text{OC}(\text{CH}_2)_{16}\text{CH}_3$$

20. aldehydes and ketones

22. Hydrolysis is the reaction of a compound with water in which the compound is broken down into simpler compounds, such as the hydrolysis of an ester into an alcohol and a carboxylic acid. Hydrogenation is the addition of hydrogen to a carbon-carbon double or triple bond.

24. A detergent's solubility in water is not affected by metal ions or by changes in the pH of water, but the solubility of soap is affected by these factors.

26. Phospholipids molecules contain a polar region and a nonpolar region and are good emulsifying agents. A fat molecule is nonpolar and is not a good emulsifying agent. In fact, phospholipids are used to emulsify fats and make them water soluble.

28. phospholipids

30. Cholesterol is a steroid component of cell membranes.

32.   ester(cyclic), alcohol(2), alkene

34.   (a)   Cellular membranes provide a barrier between the aqueous solution of the
              cellular cytoplasm and the extracellular fluid. They ensure that all the
              components of the cell are kept close together.
       (b)   A nonpolar compound would most easily diffuse through the nonpolar cellular
              membrane.

36.   (In these structures the box represents the glycerol molecule)

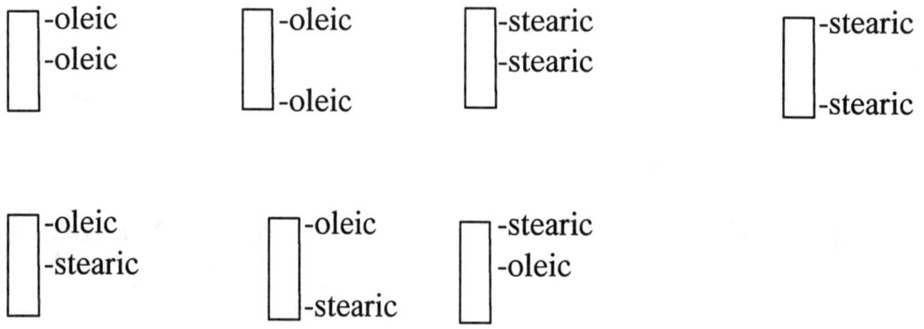

       ☐-oleic   ☐-oleic   ☐-stearic   ☐-stearic
       -oleic   -oleic   -stearic   -stearic

       ☐-oleic   ☐-oleic   ☐-stearic
       -stearic   -stearic   -oleic

38.   Naturally occurring peanut oil has only the cis arrangement around its double bonds.
       The hydrogenation process results in the production of both cis and trans
       arrangements around the double bonds of margarine molecules. The addition of trans
       isomers increases the melting point of the soybean oil.

40.   While plants produce sterols, they cannot synthesize cholesterol.

42.   Only the products of the four reactions are shown.

       (a)

       $$CH_2OC(CH_2)_{16}CH_3$$

       $$+ H_2 \longrightarrow CHOC(CH_2)_{16}CH_3$$

       $$CH_2OC(CH_2)_{16}CH_3$$

304

(b)

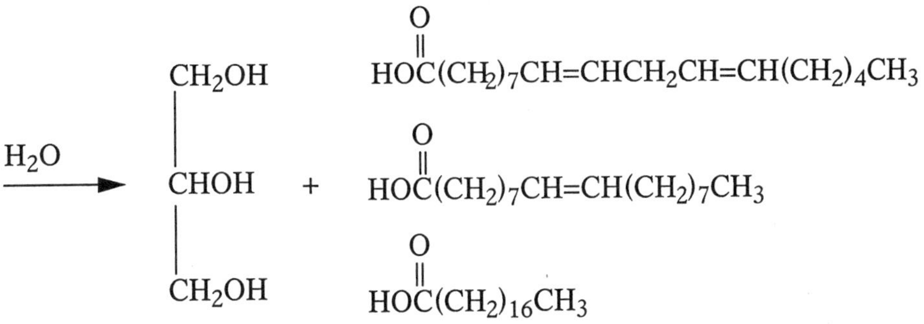

$$\xrightarrow{H_2O}$$

$\begin{array}{l} CH_2OH \\ | \\ CHOH \\ | \\ CH_2OH \end{array}$ + $\begin{array}{l} \overset{\displaystyle O}{\overset{\|}{HOC}}(CH_2)_7CH=CHCH_2CH=CH(CH_2)_4CH_3 \\[6pt] \overset{\displaystyle O}{\overset{\|}{HOC}}(CH_2)_7CH=CH(CH_2)_7CH_3 \\[6pt] \overset{\displaystyle O}{\overset{\|}{HOC}}(CH_2)_{16}CH_3 \end{array}$

(c)

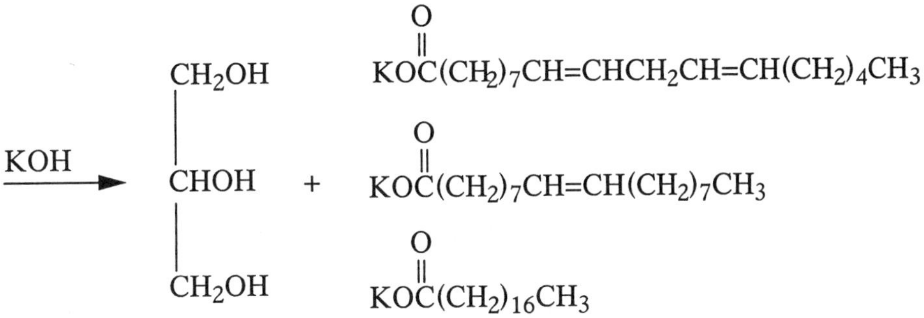

$$\xrightarrow{KOH}$$

$\begin{array}{l} CH_2OH \\ | \\ CHOH \\ | \\ CH_2OH \end{array}$ + $\begin{array}{l} \overset{\displaystyle O}{\overset{\|}{KOC}}(CH_2)_7CH=CHCH_2CH=CH(CH_2)_4CH_3 \\[6pt] \overset{\displaystyle O}{\overset{\|}{KOC}}(CH_2)_7CH=CH(CH_2)_7CH_3 \\[6pt] \overset{\displaystyle O}{\overset{\|}{KOC}}(CH_2)_{16}CH_3 \end{array}$

(d)

$$\xrightarrow{Br_2}$$

$\begin{array}{l} \overset{\displaystyle O}{\overset{\|}{CH_2OC}}(CH_2)_7CHBrCHBrCH_2CHBrCHBr(CH_2)_4CH_3 \\[6pt] | \\ \overset{\displaystyle O}{\overset{\|}{CHOC}}(CH_2)_7CHBrCHBr(CH_2)_7CH_3 \\[6pt] | \\ \overset{\displaystyle O}{\overset{\|}{CH_2OC}}(CH_2)_{16}CH_3 \end{array}$

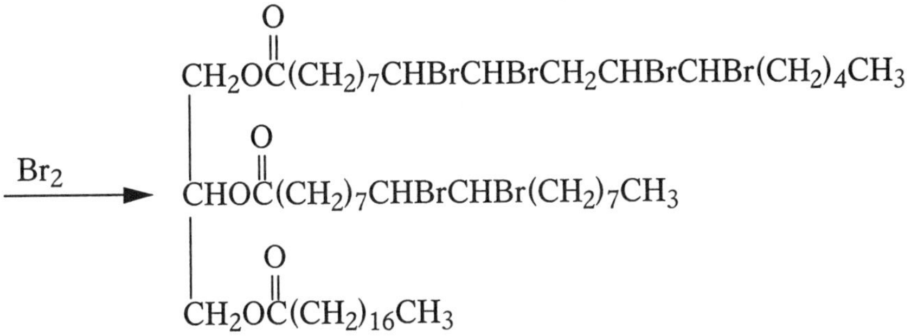

Chapter 17

2. (a) glycine   (b) alanine   (c) leucine   (d) valine

4. (a)

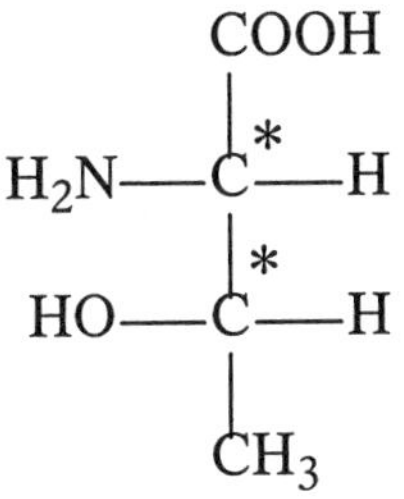

$\begin{array}{c} COOH \\ | \\ H_2N-\overset{*}{C}-H \\ | \\ HO-\overset{*}{C}-H \\ | \\ CH_3 \end{array}$

(b)

$\begin{array}{c} COOH \\ | \\ H_2N-\overset{*}{C}-H \\ | \\ H-\overset{*}{C}-CH_3 \\ | \\ CH_2CH_3 \end{array}$

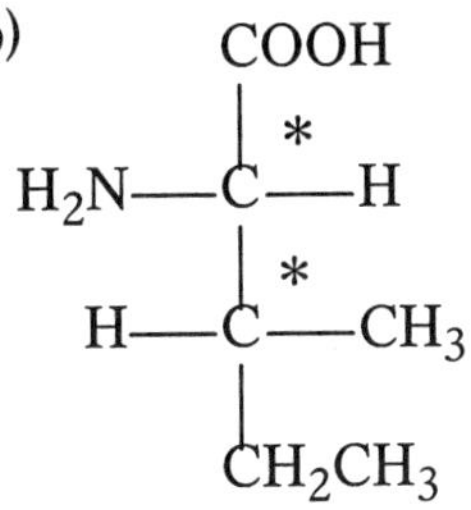

6. No, rice contains insufficient amounts of the essential amino acids lysine and threonine.

8.

(a)
$$CH_3-\overset{\overset{\displaystyle CH_3}{|}}{C}H-\overset{\overset{\displaystyle \overset{+}{N}H_3}{|}}{C}H-\overset{\overset{\displaystyle O}{\|}}{C}-O^-$$

(c)
$$\langle\bigcirc\rangle-CH_2-\overset{\overset{\displaystyle \overset{+}{N}H_3}{|}}{C}H-\overset{\overset{\displaystyle O}{\|}}{C}-O^-$$

(b)
$$CH_3-\overset{\overset{\displaystyle OH}{|}}{C}H-\overset{\overset{\displaystyle \overset{+}{N}H_3}{|}}{C}H-\overset{\overset{\displaystyle O}{\|}}{C}-O^-$$

(d)
$$HO-\langle\bigcirc\rangle-CH_2-\overset{\overset{\displaystyle \overset{+}{N}H_3}{|}}{C}H-\overset{\overset{\displaystyle O}{\|}}{C}-O^-$$

10. Blood proteins are amphoteric; they can act both as acids and bases in neutralizing additions of acid or base to the blood, and buffering the blood against changes in pH.

12. At their isoelectric points protein molecules will be electrically neutral and tend to cluster together and precipitate out of solution.

14.

(a) pH = 6.0
$$H_3N^+\text{-}CH\text{-}\overset{\overset{\displaystyle O}{\|}}{C}\text{-}O^-$$
$$|$$
$$CH_2$$
$$|$$
$$CH(CH_3)_2$$

(b) As the pH increases leucine molecules will become negative, repelling one another. The negative leucine molecules will attract water molecules and become more soluble.

$$H_3N^+\text{-}CH\text{-}\overset{\overset{\displaystyle O}{\|}}{C}\text{-}O^- + OH^- \longrightarrow H_2N\text{-}CH\text{-}\overset{\overset{\displaystyle O}{\|}}{C}\text{-}O^- + H_2O$$
$$\qquad | \qquad\qquad\qquad\qquad\qquad\qquad |$$
$$CH_2CH(CH_3)_2 \qquad\qquad\qquad\qquad CH_2CH(CH_3)_2$$

(c) As the pH is decreased, leucine molecules will take on a net positive charge and become more soluble in water than the neutral molecule.

306

$$O \qquad\qquad\qquad\qquad O$$
$$\parallel \qquad\qquad\qquad\qquad\qquad \parallel$$
$$H_3N^+\text{-CH-C-O}^- + H^+ \longrightarrow H_3N^+\text{-CH-C-OH}$$
$$| \qquad\qquad\qquad\qquad\qquad\qquad |$$
$$CH_2CH(CH_3)_2 \qquad\qquad\qquad\quad CH_2CH(CH_3)_2$$

16. Proteins act as catalysts and they are structural materials in the body. They make up the contractile system of muscles. They act as antibodies and as hormones regulate the body's glandular activities. In the blood, they maintain fluid balance, transport oxygen and carbon dioxide and they are part of the clotting mechanism.

18. Phe-Ser-Ile   Ser-Phe-Ile
    Phe-Ile-Ser   Ile-Phe-Ser
    Ser-Ile-Phe   Ile-Ser-Phe

20.

H_3N^+-CH-$\overset{O}{\overset{\parallel}{C}}$-NH-CH-$\overset{O}{\overset{\parallel}{C}}$-NH-CH-$\overset{O}{\overset{\parallel}{C}}$-NH-CH-$\overset{O}{\overset{\parallel}{C}}$-NH-CH-$\overset{O}{\overset{\parallel}{C}}$-O⁻

CH₂   H   CH₂   CHCH₃   CH₂

OH   (benzene ring with OH)   CH₂   SH

CH₃

N-terminal end   OH   C-terminal end

22. Phe-Val-Asn-Gln-Tyr-Asp

24. (a)   hydrogen bonding
    (b)   hydrogen bonding, disulfide bridges, salt bridges, or hydrophobic interactions
    (c)   hydrogen bonding
    (d)   hydrogen bonding

26. Silk would be more disrupted by heat than would keratin because the polypeptide chains in silk are held together only by hydrogen bonds, whereas keratin's structure is held together by both hydrogen bonds and disulfide bridges.

28. The R-group of cysteine contains a disulfide group that is easily oxidized to form a disulfide bond that stabilizes the native state.

30. Hydrogen bonding, hydrophobic interactions and salt bridges

32. $-CH_2CH_2CH_2CH_2NH_2 + H^+ \rightleftharpoons -CH_2CH_2CH_2CH_2NH_3^+$

34. A diet consisting mainly of corn could be supplemented with soybeans or milk to supply the missing amino acids lysine and tryptophan.

36. Albumin has a negative charge at physiologic pH.

38. Tannic acid precipitates proteins on the surface of the wound producing a protective coating.

40. The more concentrated alcohol solution will rapidly precipitate proteins on the surface producing a barrier and protecting the interior of the bacteria. The less concentrated alcohol solution will diffuse into the bacteria cell, denaturing the proteins in the interior of the cell and killing the bacteria.

Chapter 18

2. (a) Catabolic reactions are reactions which produce cellular energy by breaking down molecules.
   (b) Anabolic reactions are biosynthetic reactions.

4 Because an enzyme is not consumed in the reaction it catalyzes, it can be used over and over again.

6. (a) lipids   (e) lactose
   (b) sucrose   (f) compounds with glycosidiclinkages
   (c) cellulose   (g) proteins
   (d) peptides   (h) esters

8. (a) ligase   (b) isomerase   (c) hydrolase

10. 30,000,000 molecules

12. The induced fit theory explains the action of some enzymes by stating that the substrate, as it is drawn to the enzyme, induces the flexible enzyme to take on a shape that matches the substrate.

14. Boiling water will denature the protein in microorganisms, thereby killing them and

sterilizing whatever is boiled in the water.

16. Lead ions bind irreversibly to the sulfhydryl groups on enzyme molecules, inactivating them. This disrupts normal metabolism and causes the symptoms of lead poisoning.

18. Methotrexate resembles the vitamin folic acid and can compete with folic acid. This ties up a pathway for the synthesis of DNA in rapidly dividing cancer cells.

20. An increase in isoleucine concentration decreases the rate of conversion of threonine to α-ketobutyric acid because isoleucine inhibits the action of the enzyme threonine deaminase.

22. (a) Prostaglandins raise and lower blood pressure, regulate gastric secretions, cause uterine contractions, and dilate the opening of the air passages to the lungs. They also produce fever and inflammation.
    (b) Prostaglandins are fatty acids produced in most cells and, like cyclic AMP, they carry out the messages that the cells receive from hormones.

24. Vitamins are classified based on their solubility. They are either soluble in water or in fat.

26. Coenzyme A

28. To prevent the loss of the vitamins in the water and the destruction of the vitamin's activity by the heat.

30. (a) Vitamins A and D are fat soluble vitamins and are stored in body fat, but vitamins C and B_1 are water soluble vitamins and are readily excreted by the body.
    (b) Both a deficiency and an excess of vitamin D cause a disruption in calcium metabolism. A deficiency will cause rickets, and an excess can cause growth retardation. Because excess vitamin D is stored in the fat and not readily excreted, a toxic effect can readily occur.

32. vitamin K

34. The enzymes used in detergents were both fat-cleaving (lipases) and protein-cleaving (proteases) enzymes. They digested the fats and proteins found in most stains.

36. The enzymes are in inactive forms and must be activated to begin the formation of blood clots.

38. Boiling in water will remove the water soluble vitamins from broccoli, and long exposure to heat on the steam table will destroy much of the vitamins A and C that remain.

40. p-Aminobenzoic acid is not required for growth by animal cells. Therefore, sulfa drugs are toxic only to bacteria.

42. Vitamin A is a fat-soluble vitamin. Excess vitamin A is stored in the liver and will be released as the body needs it.

Integrated problems

2. Osmotic pressure depends on the number of particles in the cell. Maltose contains two glucose molecules. A starch molecule can contain thousands of glucose molecules. It would take several thousand maltose molecules to store the same amount of glucose as one starch molecule.

4.

6.

$$\underset{\overset{\displaystyle CH_2}{\underset{\overset{\displaystyle C=O}{\underset{\displaystyle OH}{|}}}{|}}}{H_2N-CH}-\overset{\overset{\displaystyle O}{\|}}{C}-NH-\underset{\overset{\displaystyle CH_2}{|}}{CH}-\overset{\overset{\displaystyle O}{\|}}{C}-OCH_3$$

8. (a)

(b) Somatostatin would attach to receptor sits on the surface of cells in the pancreas signaling the cell to stop the synthesis and release of insulin.

10. (a) Beriberi results from a lack of thiamine, vitamin B_1, in the diet.
(b) The thiamine present in the rice was being removed with the outer coating of the rice, and a dietary deficiency was produced by the diet of polished rice.
(c) The rice was enriched with thiamine.

Chapter 19

2. (a)

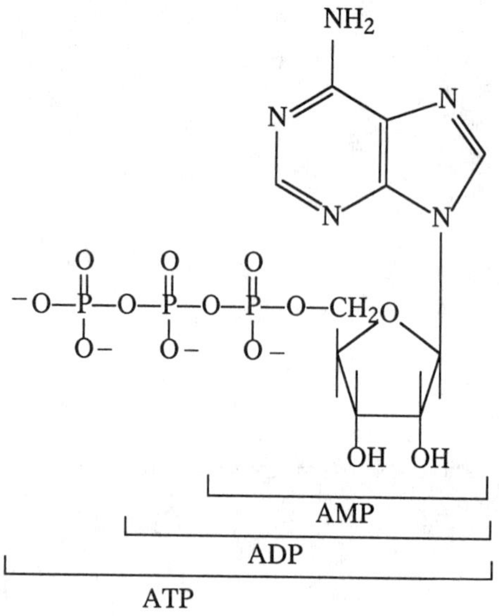

4. ATP molecules supply the energy required for the anabolic reactions occurring within the cell.

6. bile salts, bile pigments, phospholipids, and cholesterol

8. glycerol, fatty acids, monoacylglycerols, and diacylglycerols

10. Table 19.1 summarizes the digestion of carbohydrates, lipids and proteins.

12. A diet for a person suffering from hypoglycemia should be high in protein and should contain a moderate amount of fat and carbohydrate. No meal or snack should be high in carbohydrate to avoid an overstimulation of the pancreas.

14. (a) glycolysis

(b) glucose + 2ADP + P$_i$ $\longrightarrow$ 2 lactic acid + 2ATP + 2H$_2$O

(c) The anaerobic stage occurs in the cytoplasm, and the aerobic stage in the mitochondria.

(d) aerobic stage

(e) The two series of reactions are the citric acid cycle and the electron transport chain. An acetyl CoA enters the citric acid cycle and is oxidized to two molecules of carbon dioxide and four pairs of hydrogens that are bound to hydrogen carriers. These hydrogens then enter the electron transport chain, which produces water and energy in the form of ATP.

16. The citric acid cycle produces carbon dioxide, one high energy phosphate bond in the form of GTP (which transfers a phosphate to ADP producing ATP) and four pairs of hydrogen atoms, three pairs of which are captured by NAD and one pair is captured by FAD.

18. (a) step 2 (c) step 2
    (b) steps 3, 4, 6, 8 (d) step 3 and in the conversion of pyruvate to Acetyl-CoA

20. They combine with oxygen to form water.

22. (a)

$$\frac{300 \ \cancel{kcal}}{1 \ hr} \times \frac{1 \ mol}{7.3 \ \cancel{kcal}} = \frac{41 \ mol \ ATP}{1 \ hr}$$

(b)

glucose $\longrightarrow$ 2 pyruvate + 2ATP

2 pyruvate $\longrightarrow$ 2 acetyl CoA + 2NADH + H$^+$(6 ATP)

2 acetyl CoA $\longrightarrow$ 4 CO$_2$ + 24 ATP

oxidation of 1 mol of glucose produces 32 mol ATP

$$\frac{41 \ \cancel{mol \ ATP}}{1 \ hr} \times \frac{1 \ \cancel{mol \ glucose}}{32 \ \cancel{mol \ ATP}} \times \frac{180 \ g}{1 \ \cancel{mol \ glucose}} = \frac{230 \ g}{1 \ hr}$$

24. Nausea, headache, lack of appetite, and difficult, painful breathing.

26. In yeast cells glycolysis results in the formation of ethanol and carbon dioxide while in animal cells, lactic acid and water are formed.

28. Stored fat is an energy reserve, a support and a shock absorber for the internal organs and it is a heat insulator.

30. The fatty acid is joined with coenzyme A.

32. (a)

$$CH_3CH_2CH_2CH_2CH_2CH_2CH_2CH_2CH_2CH_2CH_2\overset{\overset{\textstyle O}{\|}}{C}OH$$

| 5 | 4 | 3 | 2 | 1 |

	5 FADH$_2$	10 ATP
	5 NADH	15 ATP
	6 acetyl CoA	72 ATP
		97 ATP - 1 ATP = 96 ATP

(b) 147 ATP   (c) 164 ATP

34.

36. (a) Ketosis occurs when the level of ketone bodies in the blood exceeds the amount that can be used by the tissues and is, therefore, excreted in the urine.

(b) Liver damage, diabetes mellitus, starvation, or diets can cause a decrease in glucose metabolism and an increase in fat metabolism with the accompanying increase in ketone bodies.

(c) Ketosis causes acidosis.

38.

40. The -keto acids produced in oxidative deamination of amino acids can be oxidized in the citric acids cycle, they can be converted to other amino acids by transamination and they can be used to synthesize lipids and carbohydrates.

42. Uremia. The symptoms are: nausea, irritability, drowsiness, elevated blood pressure and possibly anemia.

44. Glucose is used first, then glycogen is broken down and used.

46. Excesss acetyl-CoA is converted to the ketone bodies.

48. Each time a hydrocarbon is oxidized energy is released. Fat molecules are much lower oxidation state than carbohydrates and therefore contain more chemical energy.

50. The fasting person will be metabolizing tissue proteins, and water is necessary for oxidative deamination, for the citric acid cycle, and in urine formation.

52. Patient A may be suffering from diabetes mellitus. The normal blood glucose level should be 60-100 mg/100 mL. Patient A's blood glucose began at a higher than normal level. The large increase in blood glucose after drinking the liquid should have caused an increase in production of insulin and a rapid decrease in blood glucose as shown in the graph for patient B.

Chapter 20

2. The Central Dogma of Molecular Biology states that DNA directs the making of its own copy and the transfer of the genetic information to RNA. RNA directs the transfer of this information to the amino acid chain of a protein.

4. A nucleotide is composed of a nitrogen-containing base, a five-carbon sugar, and a phosphate group.

   A nucleic acid is a polymer of many nucleotides.

6. (a)   (b)

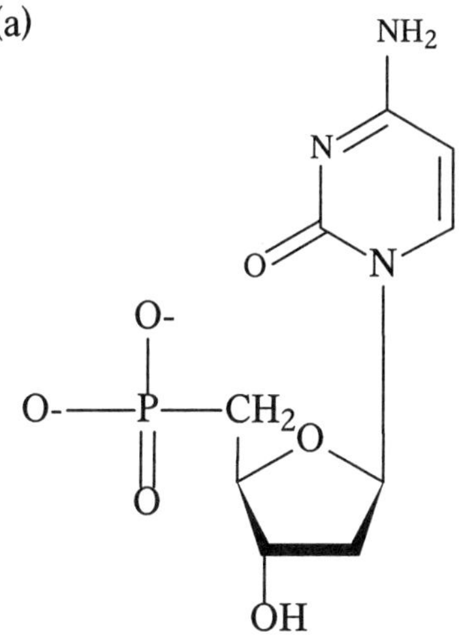

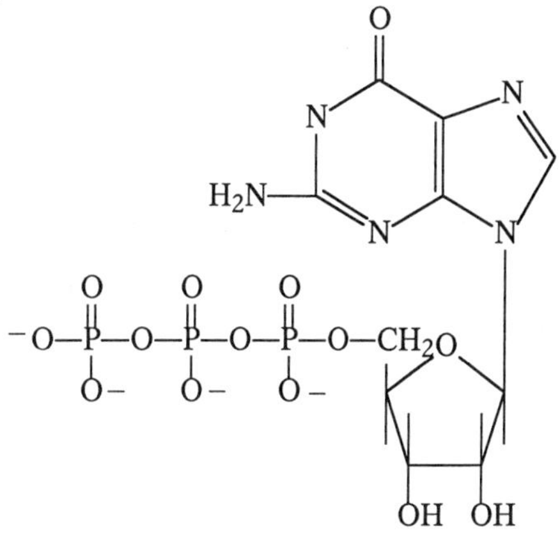

315

8.

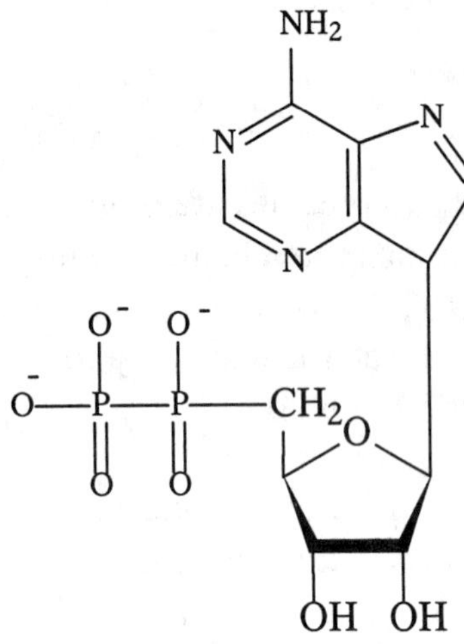

NH$_2$

adenosine diphosphate

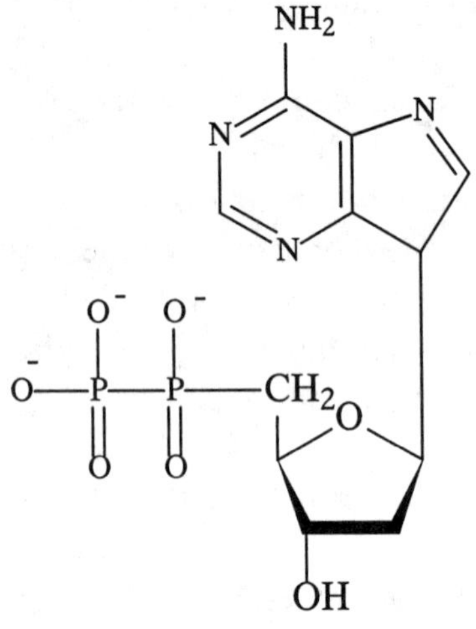

NH$_2$

deoxyadenosine diphosphat

10. Thymine and guanine both have carbonyl groups on the first position. A carbonyl group cannot hydrogen bond with another carbonyl group. Also thymine and guanine have an amino group in the middle of the three positions that form hydrogen bonds. An amino group cannot form hydrogen bonds with another amino group. So thymine and guanine cannot form hydrogen bonds with one another in either the first or second position that form hydrogen bonding in DNA.

12. For a DNA molecule to replicate, the two strands of the helix unwind, and each strand serves as a template for the synthesis of a new strand of DNA, forming two daughter DNA molecules each having an original DNA strand and one newly-made strand.

14. The nucleotides of DNA contain the sugar deoxyribose and the base thymine, and the nucleotides of RNA contain the sugar ribose and the base uracil in the place of the base thymine. The structure of DNA is a double stranded helix. RNA molecules exist as single polynucleotide strands.

16. (a) TAC GGT GTA CAT AAC TTG GGG TAA GAC ACT
(b) Met-Pro-His-Val-Asn-Pro-Pro-Ile-Leu

18. (a) hnRNA is a complete transcription of the entire DNA base sequence. mRNA contains only the sequence of bases that codes for the amino acids of the protein being synthesized.
(b) Special enzyme complexes of DNA and protein cut and splice the hnRNA in a series of reactions that producees mRNA.

20. The sequence of steps is the same except that the amino acids coded for are different.

22. Met-Thr-Lys-Arg-Arg-Gly-Ala-Ala-Ser (stop)

24. The piece of DNA that would code for cortisone is synthesized or isolated from human DNA. It is inserted into a plasmid which is then inserted into E. coli. The E. coli are grown in large vats and produce the cortisone along with bacterial proteins. The cortisone is then isolated and purified for use.

26. Examples are: cystic fibrosis, albinism, hemophilia, thalassemia, galactosemia, color blindness, and phenylketonuria.

28. (a) The effects of PKU can be minimized by feeding the infant special formula that doesn't contain phenylalanine.
    (b) Possibly, if the DNA that codes for phenylalanine hydroxylase could be inserted into human liver cells that then could be injected into the patient's liver.

30. (a) ATGTGTTACATTCAAAACTGCCCCCTGGGATAG
    (b) AUG UGU UAC AUU CAA AAC UGC CCC CUG GGA UAG
    (c) UAC ACA AUG
    (d) Cys-Tyr-Ile-Gln-Asn-Cys-Pro-Leu-Gly

32. The insulin produce by recombinant DNA is human insulin so people should not be allergic to it even if they are allergic to bovine insulin.

34. A mutation in RNA will produce defective proteins that can cause the death of one cell, but these mutations will not be passed from cell to cell and affect the entire organism.

Chapter 21

2. Because interior cells are too far from the surface.

4. blood, urine, lymph, digestive juices, synovial fluid, interstitial fluid

6. albumins, globins, fibrinogen

8. to reseal broken blood vessels or the blood will all leak out of the circulatory system

10. leukocytes

12. It is in the form of bicarbonate.

14. An increase in pH corresponds to a decrease in the concentration of protons. Equilibrium I will shift to the left causing an increase in the blood's ability to transport oxygen.

16. The shift to the left of equilibrium II due to high carbon dioxide partial pressure, increases the concentration of protons. This increase in proton concentration (decrease in pH) will cause equilibrium I to shift to the left.

18. A hypertonic solution has a higher osmotic pressure than the solution to which it is compared. A solution that is hypotonic has a lower osmotic pressure than the refernce solution.

20. The net flow of solvent is from the solution with lower osmotic pressure to the solution with higher osmotic pressure.

22. In cardiac failure, the heart does not efficiently pump the blood from the veins. The blood pressure in the veins increases. This increase in venous blood pressure reduces the flow of fluid from the tissues into the blood vessels. Fluid builds up in the tissue. This is edema.

24. Adding acid:
$$H_2PO_4^- \longleftarrow HPO_4^{2-} + H^+$$
Adding base:
$$H_2PO_4^- \longrightarrow HPO_4^{2-} + H^+$$

26. Because calcium is needed for the synthesis of thrombin.

28. The conversion of fibrinogen to fibrin.

30. The higher the concentration of dissolved solids in the urine, the higher the specific gravity.

32. sweating, fever, vomiting, diarrhea

34. Because of the dissolved glucose and other compounds.

36. Meals that are high in fruits and vegetables and some urinary tract infections may cause high urine pH.

38. The deamination of amino acids in the kidneys.

40. High blood pH corresponds to low proton concentration. The phosphate buffer system will shift to release protons. The ratio $HPO_4^{2-}/H_2PO_4^-$ will increase if blood pH increases.

42. Urea, uric acid, creatinine. Urea is the main nitrogen-containing waste product of urine.

44. Because the amount of urochrome produced by the body each day is almost constant, the concentration of this compound will decrease as the urine volume increases. As the concentration of urochrome decreases the color of urine becomes lighter yellow.

46. Because of their large size, proteins do not normally pass through the membranes in the kidney and therefore they should not normally be present in the urine.

48. Higher oxygen partial pressure will shift equilibrium I in section 21.4 to the right. This increases the concentration of hydrogen ions. The pH decreases.

50. $2.25 \text{ g} \times \dfrac{1000 \text{ mg}}{1 \text{ g}} = 2250 \text{ mg}$

$200 \text{ lb} \times \dfrac{1 \text{ kg}}{2.2 \text{ lb}} = 90.9 \text{ kg}$

$\text{creatinine coefficient} = \dfrac{2250 \text{ mg}}{90.9 \text{ kg}} = 24.8 \text{ mg/kg}$

A creatinine coefficient of 24.8 mg/kg is in the high range for men. This indicates a high percentage of muscle tissue. This person is definitely not a couch potato.

52. Lactose is present in urine during the late stages of pregnancy and during lactation. Lactose can be distinguished from glucose in the urine by a fermentation test. Glucose ferments, but lactose does not.

Integrated Problems

2. One of the effects of epinephrine is to cause a sharp increase in the rate of glycogenolysis in the liver and therefore, increase the amount of glucose in the blood.

4.   Compactin will competitively inhibit HMG CoA reductase thereby inhibiting the synthesis of cholesterol by the cells. Cells will have to increase cholesterol (LDL) uptake by increasing the number of LDL receptors on the cell surface.

6.   The codon UAG at position 17 would order the protein synthesis in the ribosomes to stop at position 17. Since the normal beta-chain of hemoglobin is 146 amino acids long, stopping the synthesis at position 17 would cause a severely altered hemoglobin to be synthesized.

NOTES

NOTES

NOTES

NOTES

NOTES

NOTES

NOTES